Earth education... a new beginning

by

STEVE VAN MATRE

Illustrated by Jan Muir

Cover design and book layout by Ben Bragonier

Design assistance by Jim Wells, Matt Haag and
Dave Wampler

Typesetting by Jeff Levin of Pendragon Graphics

First Printing, April 1990

Printed on recycled paper.

ISBN: 0-917011-02-3

Library of Congress: 89-085855

Published by
The Institute for Earth Education
Box 288
Warrenville, Illinois 60555
U.S.A.

"Whatever you can do, or dream you can, begin it. Boldness has genius, power and magic in it."

— Johann Goethe

⊕

"I believe that we should read only those books that bite and sting us. If a book we are reading does not rouse us with a blow on the head, then why read it?"

— Franz Kafka

⊕

"This is a present from a small distant world.   . . . We are attempting to survive our time so we may live into yours."

— Recorded Message on board the
Voyager space probe

Dedicated
to
the worldwide volunteer staff associates
of
The Institute for Earth Education

tABLE OF CONTENTS

Chapter Three

Chapter Four

bORDER nOTES

aCKNOWLEDGMENTS

Naturally, for a book like this there would be a lifetime of people to thank, so I hope I will be forgiven if I only include here a highly abbreviated version of the entire list.

Several long-time Associates of the institute reviewed the manuscript and made numerous helpful suggestions: Bruce Johnson, Laurie Farber, David Siegenthaler, Ginger Wallis, Eddie Soloway, and Donn Edwards.

Dave Wampler coordinated the final production details, Tim Bird typed in all the corrections and changes on the word processor, Benjamin Bragonier served as our overall art director, and Jan Muir added the wonderful drawings.

Although I had been working on ideas for the book, when time permitted, for several years, Jim Wells, Kirk Hoessle, Eddie Soloway and Dave Wampler joined me for an important planning session in 1984, and fortunately, Jim came back at the end to help out with the book's final layout and design.

Pat Walkup, Harry Hoogesteger, Bill McKinney, John Clarke, Bruce Elkin, and Helene Phelps, although no longer directly involved as part of the institute's staff, contributed to the development of these ideas in numerous ways even when they may not have realized they were doing so.

Of course, my graduate students at George Williams College and Aurora University have given me the chance to refine my points year after year, and their questions and observations are reflected in much of our work.

Finally, I would like to thank Martha and Bill. Although they are fictitious characters that I refer to in my workshops (or in our journal, "Talking Leaves"), their comments are based on actual conversations and correspondence over the years. Even though I have been somewhat less than charitable at times in my retorts to their queries, they have prodded all of us into thinking about things that needed further examination and clarification and that is always of value. So to all the Marthas and Bills out there, thanks for keeping us on our toes even when I didn't relish the opportunity or respond as kindly as I should have.

CREDITS

We gratefully acknowledge the following authors and publishers for permission to use their writings:

<u>Deep Ecology: Living as if Nature Mattered</u> by Bill Devall and George Sessions (Gibbs Smith, Publisher, 1987)

Kelley, <u>The Home Planet</u>, © 1988 by Kevin W. Kelley. Reprinted with permission of Addison Wesley Publishing Company, Inc., Reading, Massachusetts.

<u>Out of This World</u>, by Michael Page and Robert Ingpen, Lansdowne Press, New South Wales, Australia. © copyright text Michael Page 1986. Reprinted with permission of the author.

Note: Omissions brought to our attention will be credited in subsequent printings.

An ancient symbol
representing the
earth

iNTRODUCTION

For the past twenty years we have been led to believe that there is a significant educational response underway around the world for dealing with the environmental problems of the earth. It is not true. Right from the start that effort has been co-opted, and diluted, and trivialized. Now we are on the verge of losing a whole generation of planetary citizens who could have set the course for a different future. There is not much time left. This book is about a new path that the world's largest group of professional educators in the environmental field has already taken. We hope you will join us.

I am afraid parts of our account are not going to be much fun to read. I have agonized over my approach here for some time, and I have listened to all manner of contrary advice. A couple of my reviewers even said it would be a great work if I just left out the opening chapter. But I am convinced that environmental education has become little more than so much mush to so many people that if I don't make our case strongly enough, folks will just stir us in with all the other educational supplements available. Besides, as you will see, we no longer believe we are environmental education; we think we are an alternative to it.

Why do I sound so angry? Sure, it bugs me that I have watched a number of people sacrifice significant portions of their lives to the goal of making a focused educational response to our environmental crisis, only to be thwarted over and over by people who are patently less concerned, less committed, and less skillful than they are. However, the main source of my vitriolic comments is the escalating destruction of this planet and a feeling of immense sorrow that we have failed to do enough. In a couple of generations we have destroyed much of the earth; in a couple of more it may be too late to reverse the process. I say a pox on all those in education who feel no bitterness, who refuse to speak out, who cannot bring themselves to take a negative stand against the sad state of our field.

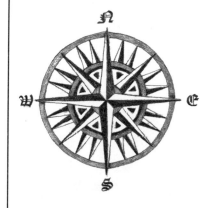

A SMALL SUGGESTION...

Reading this book may turn out to be a bit of a journey itself. Like many new experiences, the first chapters could provide some tough going, but please forge ahead. There are lots of good things coming, and once you get your bearings and get into the rhythm, things should become easier.

Frankly, if you are the kind of a traveler though, who likes to have the experience first, then read the guidebook, or if you prefer getting the positive before the negative, then you may want to read the last chapter of this book first.

Although such words may not make it appear so, I have actually tried to drain away as much of the accumulated resentment as possible from these pages. I have soaked the words in the waters of both time and distance and poured off the resulting liquid on several occasions. In the end, I am afraid it is still some pretty strong stuff. I would make just one request of you: don't let my histrionics in such places get in the way of a thoughtful analysis of my points. I may be abrasive, but it is not because I love the earth any less. It may just be that I love it more.

As far as I know, most of the folks I complain about in the sickening state of environmental education depicted here are all good people, committed to the earth, and, no doubt, they have also sacrificed in pursuit of their beliefs. But one can still believe that many of them were dead wrong in how they went about doing it, while others simply didn't do enough.

In the Institute's Members Survey last year, someone asked if it was really necessary for us to come across so aggressively, almost like we were in a battle. But that is exactly the point. We are. We are in a battle for the hearts and minds of those who can make a significant difference in the health of the earth. Should we not fight for the attention of those teachers and leaders who may yet be the best hope for the future? I realize some of them would follow the path we have taken more readily if I took a less antagonistic approach here, but this way, when people join us, they will be clear about where they are going.

You see, we don't think just as long as everyone espouses the same general goals that everything will be fine. We feel the situation is so crisis-oriented that it will require much more from us than just a breathless, "Oh, we're all in this together, just working away." I am reminded of one of those folks who claim we should be building bridges instead of cutting others down, who went off to a centre to put together a new environmental education program for them, and after spending a single weekend patching together a potpourri of various activities from other people's work, returned saying, "Oh, I just love putting together curriculums." It is going to take a lot more than that, believe me.

There is little doubt that our environmental crisis will worsen in the years ahead. Will we be ready on

"Having to squeeze the last drop of utility out of the land has the same desperate finality as having to chop up the furniture to keep warm."

— Aldo Leopold

the educational front to respond to that situation, or will we still be mouthing organizational platitudes and mixing educational potpourris while the other creatures of the earth, as well as many of our fellow human passengers, sicken and die around us?

People tell me that we should be more positive in our approach, but hey, we are not selling aspirin here. This is not just a matter of being neutral or mildly disapproving about the other group's product. We believe the other product misleads people into thinking it represents a cure when it doesn't. In fact, we are afraid it may be contributing to the disease instead of arresting it. (By leading people to believe they are dealing with the problems when in reality they may be subtly conveying the very ideas that caused them.) Please explain to me how you can take a positive approach in a situation you believe is that grave.

If you are a newcomer to this field, now is the time to make a choice. The path for an educational response to our environmental crisis divides here, and this book will attempt to explain a route you may not have known existed. If you are an old-timer who has been pursuing this route all along on your own, then after reading these pages, we hope you will stand up and let others know there is a genuine alternative to environmental miseducation. Give them a headstart on their journey. Don't let "cornucopian" thinking dominate our organizations and publications and conventions. Of course, if you have already started down the other path, we hope you will hear our cries of alarm and retrace your steps.

To be honest, the earth needs your leadership like never before. You will read in the following pages that the earth is in trouble and we are the problem. But that can also be turned around the other way: the earth is in trouble and we are the solution.

As I tell my graduate students each fall, as far as I can foresee, there is no pursuit, no profession, no position out there of any more importance than the journey they are undertaking. What we are talking about here is literally the health of the planet earth.

It is a journey I hope you will undertake too. We may never meet, but please consider joining us on the path of earth education. It is a demanding one. It

"To every man is given the key to the gates of heaven and the same key opens the gates of hell."

— Chinese Proverb

requires passion and perseverance. It can be lonely, and no doubt you will suffer the slings and arrows of the complacent and counterproductive, but it is a path with heart.

Good journeying,

S.V.M.
Illinois Prairie Path
October, 1989

"I slept and dreamt that life was pleasure, I woke and saw that life was service, I served and discovered that service was pleasure."

— Rabindranath Tagore

PROLOGUE

Dear Reader,

Before you begin your journey in discovering the earth education path, I would like to introduce you to some special characters that will help you along the way. They are known as the Tellurian Gnomes.

In their marvelous book, Out of This World, Michael Page and Robert Ingpen gave us a general description of those magical earth spirits called gnomes:

> The international family of gnomes dates back to the era when the form of the globe consolidated out of Chaos, and the forces responsible for precious and base metals and precious stones implanted them beneath the surface of the earth. Unlike men, gnomes learn from the past and they also have the ability to predict and learn from the future. Their name derives from the Greek word gignosko, meaning "to learn, understand," and the principal gnome characteristic is an acute understanding of every aspect of the Cosmos.
>
> Gnomes are about twelve centimeters tall and formed in proportion . . . . Apart from their small size, a notable difference between gnomes and humans is their expression of ageless good humour. They lack the human facility for worrying, practise therapeutic festivity, and consequently live for several hundred years.
>
> Gnome character is helpful and benign. Gignosko provides them with insight into the spirits of all animate and inanimate creatures and objects, so that they find it easy to influence and co-operate with trees, tools, animals, plants, and every other creation of the Cosmos and its inhabitants.
>
> Their diet is largely vegetarian, basically cereals and root vegetables . . . . They brew excellent ale

"Nature is loved by what is best in us."

— Ralph Waldo Emerson

*but use it in moderation except in some festivals.
Gnomes pioneered many crafts, such as weaving
and woodworking, but they have not felt disposed
to explore more complex technologies. They live
in simple comfort and avoid the problems of
industrialisation.*

*Their original duty was that of supervising
and surveying the mineral treasures of the earth.
Each group of family looked after the lode of copper,
A vein of gold, a pipe of diamonds, a seam of
coal, or some similar resource implanted underground.
Each gnome colony lived underground, close to their
particular area of responsibility. To facilitate their
work they developed the ability to move or "swim"
through the earth.*

*Their helpful nature inspired them to assist
men in discovering natural treasures, in such ways
as guiding the feet of prospectors and influencing the
science of geology, even though* gignosko *warned
them that men would use these treasures for evil as
well as good. But even their gift of prediction, or
"future mining," could not foresee the unending
extent of human avarice and its consequences.*

*As men delved deeper into the earth countless
gnomes found themselves displaced. Clumsy miners
wrecked entire colonies and the roar of explosions
made life unbearable. Some gnomes turned against
the miners . . . but the majority decided to emigrate
to the surface and begin a new existence in the light
of day.*

*The first surface gnomes (unkindly known as
"superficial gnomes" by their underground brethren)
emerged into the forests of Britain and Europe at about
the time of King Arthur. They found the dimness of
the huge forests compatible with their underground
nature, and established colonies in the root systems
of great trees.*

*As time passed, their benign character tempted
them to aid in human affairs. The little milkmaid
weary of her task would find all the milk pails filled
before she rose in the morning, the sleeping shepherd
awoke to find his flock had been rounded up for him,
and many poor tailor or shoemaker prospered when
gnome families worked night shifts in their workshops.*

Humans often glimpsed the gnomes but never succeeded in capturing one. Gignosko *always keeps gnomes a step ahead of humans and the most cunning of gnome traps never tempted a victim. Eventually humans simply accepted their presence and knew that gnomes would not harm them.*

Gnome lifestyles changed yet again when humans felled the forests just as they had looted the earth. Gnomes had to retreat further and further away from the homes of men, and they are now seen only in the few remaining forests of Britain and Europe. In other parts of the world, gnomes have adapted themselves in various ways. Some still try to assist humans by imparting some of their own gift of understanding. A few have turned mischievous, like their cousins the gremlins, and they torment humans with minor but irritating activities. Other gnome families have travelled as far as Australia in search of a new life.

The Tellurian Gnomes, while similar in many ways to other gnomes, are an unusual race who have special responsibilities for guarding the earth's natural places.

They have large eyes, ears and hands, but very small mouths, because to learn you have to see and hear and do, but not talk. In fact, Tellurian Gnomes seldom speak.

They also have rather bulbous noses because their sense of beauty is based upon smell. The highest compliment they pay each other is to say, "You smell right." And one often hears the farewell, "May you have a day of good smells."

The Tellurians wear large, oversized coats made up of dozens of pockets, each one containing some natural treasure they are seeking a home for, or some special tool for helping them experience the riches of nature (bulkier items are carried in their multicolored, drawstring gnome bags). Of course, all gnomes like to dress up, but perhaps because of their special work, the Tellurians often adopt heroic costumes.

When they are wearing their regular clothing Tellurians have hats that look like the tops of mushrooms. In fact, if they get caught out in the open, they squat down quickly, slipping their legs into the earth, so that most people believe they are just looking at another

TELLURIAN GNOME

BULBOUS NOSE
LARGE EYES
LARGE EARS
MUSHROOM CAP HAT

LARGE HANDS
OVERSIZE COAT
POCKETS WITH TOOLS

mushroom. (Whenever you see a particularly large mushroom. you should check it out, it may actually be a Tellurian Gnome hiding there.)

Fortunately, this race of gnomes appears to have an affinity for earth educators, and you will find them offering a helping hand at every turn of our path. You will be wise to keep an eye out for them because the Tellurians have an innate sense of the earth's needs.

May you have a journey of good smells,

Steve Van Matre

Steve Van Matre

CHAPTER ONE

eNVIRONMENTAL
eDUCATION...
mISSION
gONE
aSTRAY

Environmental education may well have been the most important movement this century in terms of the health of our home, the troubled planet earth. Listen to what our political leaders were saying in the United States a couple of decades ago at the beginning of that effort:

> *The Congress of the United States finds that the deterioration of the quality of the Nation's environment and of its ecological balance poses a serious threat to the strength and vitality of the people of the Nation and is in part due to poor understanding of the Nation's environment and of the need for ecological balance. . . .* *(Public Law 91-516)*

It seems pretty clear in retrospect that what people were talking about in the sixties and early seventies was that the inhabitants of the earth did not understand how life functions here (the big picture ecological systems); they did not grasp how their own lives were directly connected to and supported by those systems; and they did not understand how they were going to have to change their lifestyles in order to live more in harmony with those systems — systems which governed all life on this small self-contained vessel they shared. In short, the earth was in trouble, and we were the problem.

So what happened to our original sense of mission and purpose? The call seemed loud and clear in the beginning, but somewhere along the way it faded into a vague, almost unrecognizable whisper. If you stop people on the street today and ask them what supports life here, they will probably be unable to comprehend what you are asking. It is likely that the most you will get is a response something to the effect that the people themselves support life, or even worse, that their city does. Isn't it tragic that most people can name a few of the trees we have planted along the streets but don't understand the flow of sunlight energy in our systems of life or the interconnectedness of all living things? We've focused on the pieces of life instead of its processes. We teach people the names of some of the parts of the earth, but fail to convey how it functions as a whole.

Most people simply do not understand how much trouble we are in here, nor why. For them, energy is oil, and the middle easterners and oil companies have

OCTAVIUS PLANTA

teamed up to manufacture shortages in order to keep prices high. Cycles are Hondas or Yamahas, and diversity means tolerating your kooky neighbors. Thanks to the efforts of our mass media, people are aware of such problems as acid rain, ozone depletion, and toxic waste, but they usually don't see the connection between their own lives and the problems (and no one dares tell them). In one recent survey over 90% of the people interviewed thought that the scientists would take care of our environmental problems. Already over twenty million people on the earth die of starvation each year, yet many of our fellow citizens still believe we will establish space colonies to solve our burgeoning population problems.

People don't understand even the simplest of environmental connections: between fast-food burgers and the destruction of the world's forests, between decorative lighting or frivolous appliances and the accelerating costs of nuclear-generated electricity, between our over-packaged and over-processed foodstuffs and our growing difficulty in maintaining clean drinking water . . . or the very direct personal effect that such situations have on their own lives. That's enough; you probably get the point. In environmental education we never did the job we set out to do.

After two decades, millions of dollars, and far more words, many of the environmental education publications and projects, training sessions and manuals designed to change people's behavior have all but disappeared. What happened? I think we blew it, and it just may have been our last best shot.

"SURVEYING tHE SUPPLEMENTALISTS"

Poke your head into almost any school today and see how much real environmental education you find going on there. I don't mean a couple of supplemental activities (inside or out) led by one or two valiant teachers, or a nature bulletin board in the hallway. I mean *focused*, *sequential* instructional *programs* as a regular, integral part of the whole curriculum. Not much luck? Try the school's closets. That's where you'll probably find the most evidence. Look for the now unused books, boxes, pouches and kits that were once

Any path will do if you don't know where you are going.

common to our field, plus the obligatory mimeographed curriculum guide. Chances are good that most of it gets very little use these days. And if a few teachers do include an environmental lesson or unit, chances are good that they still do not systematically address what environmental education set out in the beginning to accomplish, i.e., how life functions ecologically, what that means for people in their own lives, and what those people are going to have to do in order to lessen their impact upon the earth.

Next, stop by your average nature centre or outdoor school and see what you find there as well. The name of the place may have changed, but the staff is probably back to identifying the plants, doing tombstone rubbings, taking Ph tests, reading the weather gauges, making maple syrup, etc. In other words, they are offering a random assemblage of outside activities (yes, with some sensory awareness experiences from Acclimatization and a few similar environmental games thrown in) all tied together by a schedule rather than a desire to achieve particular learning outcomes. Don't be misled either by the name of the centre. It may be called an environmental education centre (in the U.S. perhaps even designated as a national environmental education landmark), but take a close look at what is really going on there. Think about what you see. Are they really accomplishing their objectives? Do they claim to be doing environmental education, and, if so, are they really focusing on our individual connections with the earth's ecological systems and our personal impact upon them?

In any event, it is highly unlikely that you will find much focused, sequential, cumulative environmental education programming at such places. Oh, you may find some familiar words floating around — we were all quick to adopt the jargon early on — but very few genuine learning programs were ever developed. Listen to how the staff at one well-known nature centre explained their objectives:

> *To effectively introduce participants to the beauties and critical importance of the natural environment in their lives.*

> *To motivate and enthuse participants about the natural world around them. Hopefully, this will induce a love and caring, along with the under-*

EDUCATION VS. RECREATION...

Unfortunately, many of our nature centres view themselves primarily as outdoor recreation facilities instead of education ones. In the U.S. these places are often sponsored by park districts and other agencies who have a very limited grasp of (or interest in) their potential for initiating positive environmental change. Indeed, their governing bodies may actually oppose the very thing that many people naively believe such places have been designed to accomplish. So be careful out there. In reality, a nature centre may be little more than the sine qua non of some commissioner's master plan.

standings necessary to stimulate them to take better care of that environment for themselves and others.

To give students adventure-discovery oriented experiences which provide fun and excitement while they learn.

Sound good? Unfortunately, what they were doing to accomplish these goals was not using specific learning experiences that focused on how the systems of life here directly support us as individuals and what we must do to change the way we live. No, they were offering classes on orienteering, tree and bird identification, prairie restoration, skiing, maple sugaring, wild edibles, etc. That's right. This centre had the words for environmental education but not the tune (certainly not the harmony).

After you have made the rounds of our educational institutions, sit down and sift through some of the major so-called environmental education programs that were developed. You are in for a real surprise. You will find that several of them don't even deal with basic ecological understandings, i.e., concepts like the flow of energy, or the cycling of building materials, or the interrelationships of living things. You're asking yourself, "How can someone possibly claim to have a comprehensive environmental education program and not deal directly and effectively with the fundamental base for all life on the planet: the flow of sunlight energy?" That's a good question. What's amazing is that it has been so seldom asked in our field.

Other projects, as you will see, dealt with some of the concepts (often minor ones like stream development or beach zonation), but never attempted to clarify for their participants how their lives were connected to these processes, nor suggested that they should examine their own lifestyles in light of their new understandings.

A couple of projects included a framework, even placed ecological understandings within it, but then provided only a disjointed, random accumulation of not very stimulating activities to get the job done that they had so carefully identified in their organizational structure. As a result, you often got either the activities with no good framework, or the framework with no good activities. (Of course, the easiest thing to do in

OUR DOG IS BETTER THAN YOUR DOG...

I hope you will not conclude that our criticism of other materials in these pages is merely a childish refrain designed to bolster our own image. What we are talking about here are different "animals," with fundamentally different characteristics, not just different approaches. Would we ever use an activity from one of the "projects" in an earth education program? Sure we would; if it did the job we really wanted done, in the way we wanted to do it. (Chapter Six will explain how we would go about deciding.)

such a situation is to construct a framework that is so broad people can fit anything they want within it; thus assuring, I assume, both their success and your own.)

It is also going to be fairly obvious in your examination that for some of these projects the activities were created first and their objectives formulated later. In fact, chances are good that any time you find an activity description that claims to accomplish several objectives simultaneously, you have found an activity that was not developed with a specific learning outcome in mind. Instead, a group of people probably got together to come up with things to do, then figured out what their products were going to achieve afterwards.

You should also check out how these projects placed their activities in the various subject areas while you're at it. I imagine it went something like this — picture a group of people sitting around a table commenting upon an activity like making maple syrup: "Well, let's see, they figured out the number of buckets of sap it takes to make a jar of syrup didn't they, so it's a math activity." "OK. And they listened to the sap gurgling beneath the bark that means it's a science investigation." "Don't forget when we told them how the pioneers did it. That's social studies." "True, and they had to write a report on it when they got back, so it fits in language arts as well." Want to guess how someone would justify this as an environmental education activity to begin with? "Well, we discuss multiple forest roles with the kids while they watch the sap boiling down." R-i-g-h-t. . . .

Perhaps the most damaging development though is the assertion you will find in many of these projects that leaders should use the materials in any way they like. In other words, people should just pick and choose whatever catches their fancy, or whatever happens to fit with what they are doing at the time.

Hardly any of us questioned such practices. Very few cried out, "Hey folks, if we are serious about the task of environmental education, then it won't work just to sprinkle a couple of these supplemental activities around like so much spice over a melange of other educational pursuits. We are going to have to put together some focused, sequential programs to get the job done." Why haven't you run across this exhortation in an introduction to the materials you have been

THE LOST GUIDES...

I remember hearing about one state curriculum development conference where they started out by asking the participants to bring all the EE material they could get their hands on. When everyone got together, the organizers piled this stuff on tables around the room and asked the participants to sort through it looking for good ideas. I think the major expense for this particular charade was in renting the photocopy machine that was set up in the centre of the room because that's literally how their state curriculum guide was produced.

"If you aim at nothing, you'll hit it every time"

analyzing? Probably for much the same reason that you could go to a national conference on environmental education in those days and find everything from orienteering to acid rain on the program. (Or in other words, everything from outdoor recreation to environmental studies.) Everyone was having a good time giving birth to a new profession. It was a heady period, and nobody wanted to step on anybody else's toes. (If you are an orienteering nut, please don't get too upset with me at this point. I think it's a fine activity, really. But its goal is not to help people understand how life functions here, or what that means for them in their own lives, nor how they are going to have to begin examining their lifestyles. More on the acid rain later.)

EVALUATING EE MATERIALS

Try giving any material you analyze 100 points, then subtract up to five points for each of the following questions depending upon how well each one is addressed.

⊕ Who paid for the materials or funded the developers, and are their messages hidden in the products?

⊕ Are the leaders encouraged to build complete programs with specific learning outcomes or just to sprinkle around whatever catches their fancy?

⊕ Can the leaders see the big picture they are working on and do they know how to go about fitting all of the pieces together?

⊕ Does the introduction include a lot of lofty aspirations and high-sounding objectives, while the materials fall short in achieving those goals?

⊕ Do the learning experiences really address underlying environmental concerns, or are they superficial, external treatments requiring little or no change in their participants?

⊕ Are there specific models and schedules for the leaders to consider in how they will use the materials?

⊕ Do the activities really accomplish what they claim, or does the leader have to talk them into doing their job?

⊕ Are the activities tacked on just to get some doing in there somewhere, or are they integral parts of a learning model?

⊕ Do the activities captivate and motivate the learners, or do they involve a lot of tiring and uninspired paperwork assignments?

⊕ Do a lot of activities end up with "discussion" or "follow-up" lists that look good, but for which there is no built-in motivation or mechanism to insure that they are actually carried out?

⊕ Do the activities deal with basic ecological processes and their meaning in people's lives, or do many of them focus on secondary concepts and concerns?

⊕ When they are participating in the activities, do the learners know where they are going and why?

⊕ After they are finished with the activities, do the learners know what they can do next, and are they encouraged and supported in doing it?

⊕ Are the essential activities mostly classroom-based exercises that require little or no contact with natural places and processes?

⊕ Do the activities fire the learners up about our environmental problems but fail to address how they are part of them?

By the way, we analyzed some of our earth education materials using this checklist, and we fell short on a couple of the items ourselves. The important thing is to use these questions to evaluate what you need to do with the materials you are examining.

Before we end our survey, I hope you gathered up a bunch of flyers advertising various events in the field as you poked around in those schools and centres. Examining the announcements for many of our regional workshops and state conferences continues to provide the best empirical evidence for what environmental education actually represents for many of us, i.e., good times salt and peppered with outdoor pursuits, nature crafts, curriculum enrichment activities, awareness games, socialization techniques, and environmental action projects. I think the planning team for one state environmental education conference summed up in a single act the frequent outcome of our namby-pamby, let's be everything to everyone and stay happy approach: they turned down Edward Abbey as a free keynote speaker because he was too controversial.

"definition dementia"

Naturally, any time there is a new field developing lots of newcomers rush in and add their ideas to the coalescing movement before there is a chance to sort things out very clearly. Often this cross-fertilization is helpful, but sometimes it results in strange offspring.

One of my favorite definitions that appeared in the early days was the one that says "Environmental education is education that is in, about, or for the environment." Gosh, no wonder people went astray. Using that definition what *isn't* environmental education? I just cannot imagine what we were thinking. How would it help solve the world's environmental problems by calling almost any educational pursuit that took place anywhere an environmental education activity?

Another one of those less-than-helpful definitions went something like this: "Hey Steve, this is an environment, isn't it? I mean right here in this room; isn't this an environment, too? Well then, we're doing environmental education; we're getting people in touch with this room." Sometimes these folks would go so far as to add, "We even do your thing, Van Matre. We put blindfolds on them and have them feel the walls." Once again, environmental education was quite literally everything anyone did. I guess that made it easy on us. Since everything was in the environment, we just continued doing whatever it was we had been doing, and without batting an eye we could be an environmental educator too.

Talk about confusion. I remember one professional who said he was doing environmental education because he was working with the concept of interdependence. When we asked him what he was doing to get this concept across, he replied that he had everyone in his group try to stand at the same time on top of an old tree stump, and to do so, they had to "interdepend" in holding everyone together. Yes, that is what he said, honest, and he didn't seem to grasp at all that he was talking about sociological interdependence, not ecological interdependence. Obviously, it is pretty hard to demonstrate the latter with a group of one species huddled together on a tree stump. Unhappily, this bit of interchangeability was fairly

ENVIRONMENTAL OR ENVIRON-MENTALIZED...?

Lots of people confuse environmental education with an environmental approach to education in general. These are two different ideas. Environmentalized education is a worthy goal, but as you will see, we believe it should be the supportive rather than the primary emphasis of our work. We think we need to begin changing environmental habits now, and we need carefully-crafted, focused programs for doing that job.

widespread, and the field is filled with the peculiar hybrids it spawned.

Some years ago a recreation professor at the college where I taught asked me if people weren't in the water cycle too. It was one of those questions with an answer so obvious that you didn't want to respond because the obviousness must be hiding an unseen hook. But when I replied, albeit gingerly, that yes, people were in the water cycle too, my interrogator merely beamed and said, "Well, we're doing environmental education then. We're working with people." When I asked him if the group he was working with went home with an understanding of the water cycle, the largest physical process on earth, along with an understanding of how they were personally tied into that process, or how they were going to go about lessening their impact upon it, he changed the subject. Clearly though, he had been led to believe that if you cared about the earth yourself and were working with people, then you were doing environmental education. That approach became a big problem for a new field like ours. Those folks, however well-intentioned, went about creating collections of activities that focused primarily on getting along with each other instead of getting along with the earth.

Please don't misunderstand. Helping people become more humane, caring, etc., and helping them get in touch with the space they live in, are praiseworthy goals, but lumping them together with the goals of environmental education and serving up the results as the same thing just created confusion for all concerned.

In fact, lots of leaders write environmental education when they appear to mean good education instead. For example, "environmental education is a process which uses a wide variety of learning resources, both indoors and outdoors." That sentence could just as well be a description of good education in general.

In the Belgrade Charter, the UN document outlining the present cornucopian response to our environmental problems, most of the guiding principles set forth for environmental education programs could just as well apply to any form of education. Try dropping the word environmental from some of their descriptions and see for yourself:

UNITED NATIONS CONFERENCES...

1. The United Nations Conference on the Human Environment held in Stockholm in 1972 recommended the establishment of an international environmental education program.

2. In 1975 Unesco/UNEP, responding to the Stockholm recommendation launched the International Environmental Education Program (IEEP) and sponsored an international environmental education workshop in Belgrade, inviting experts from around the world to attend. The Belgrade Charter, adopted at the close of the conference, set forth the first principles and guidelines for the worldwide environmental education movement.

3. The first Intergovernmental Conference on Environmental Education was held in Tbilisi (U.S.S.R.) in 1977 and national delegations from 66 member states formulated 41 recommendations for implementing a worldwide environmental education program.

4. In 1987 Unesco/UNEP sponsored the Moscow Congress which produced a document titled, "International Strategy for Action in the Field of Environmental Education and Training for the 1990's."

Although one can hardly quarrel with such statements, the problem is that they are so broad that they
end up providing very little useful guidance for the
practitioners in the field.

In documents like this, and the Tbilisi Report
which followed it, the authors evidently had so many
people to satisfy that they ended up just muddying the
waters of perception for everyone. And they often
refer to the built environment or the work environment in a way that suggests more of a sociological
perspective than an ecological one. For example, the
first recommendation of the Tbilisi Report states that
environmental education should also aim to preserve
historical landmarks, works of art, monuments and
sites, etc. . . .

As for the recent Moscow report, it appears that
the primary accomplishment of the decade preceding
it was to produce yet another round of conferences
and reports, without ever seriously analyzing what
was actually happening (or more appropriately not
happening) out there in the field. To their credit the
authors of this report do include some mention of the
necessity for changing individual lifestyles (at long last),
but they cushion it with so much verbiage and offset it
with so many other "management" and "developmental"
qualifiers that it gets lost amidst their 149 other
recommendations.

On my first world speaking tour in the early
eighties, a regional environmental agency in one of
the countries I visited sponsored my "Sunship Earth"
speech about our place in space. When I arrived, my
host met me at the airport, and on the way into town

TWO VIEWS
OF THE EARTH...

Basically, cornucopian
thinking sees the world as a
horn of plenty primarily for
human use and ingenuity,
and maintains that if our
economic system is left
unfettered, it will always find
new energy and resources
to replenish the supplies
tumbling out of that horn.

Deep ecology (or Neo-
Malthusian) thinking on the
other hand views the earth
as a finite vessel of diverse
life and believes that the
human species must learn
how to curb its appetites and
work for a sustainable earth
society on behalf of all the
creatures that share the
vessel they inhabit.

The World Conservation
Strategy attempted (and
failed) to bridge the gap
between these two conflicting
views, and thus, like the
international environmental
education movement, failed
to provide a powerful,
motivating vision for our
future.

began telling me about how far they had come with environmental education in that part of the world. Frankly, I tried to stop him (I usually shy away from such conversations before a speech lest my hosts think I am attacking them when I'm on stage later), but he was on a roll, and I couldn't distract him. "We've stopped arguing about what it is," he explained, "and we're just getting on with the job of doing it." I always wondered if he understood what he was actually saying. Since they had never decided what it really was, how would they ever know when they were really doing it?

The ultimate response in the early days came however from those who wanted to end any conversation about definitions by saying: "It's just semantics." Of course, they were right, but that's exactly why it was so important. Semantics is the study of changes in the meaning of words. So instead of closing off discussion, these people should have been encouraging more of it in order that the changes in this case could be analyzed and clarified for everyone.

"iNFUSION pOLLUTION"

Another popular idea that surfaced early on in EE was also to cause endless problems. This was the position that environmental education was process not content. The advocates of this approach said we should use an environmental perspective to tie all learning together. No one ever demonstrated exactly how to do this, but it sounded good in the telling. That is why we should have been suspicious from the beginning I suspect; it sounded *too* good. In fact, becoming truly integrated in an overall approach to education has been the recurrent dream of educators for centuries. Add to this the supplementalist approach that teachers were going to "infuse" every part of the curriculum with environmental education, and you have a recipe for failure. Oh, it looked promising on the surface, but when you stop and think about it, would you want to end up in an operating room under the scalpel of a surgeon who had learned her skills in this way?

Or even better, try this: the next time you decide to fly somewhere, as you board the plane, poke your head in the cockpit and ask the pilot where he learned how to fly. If he responds, "I guess I got a little piece

AN EARLY
WARNING SIGN...

Perhaps the clearest example of our failure to sort out what we were doing (and the result of our inaction as well) was the appearance of something called energy education. If environmental education had been doing its job, what need would there have been for energy education? Think about it.

Supplemental and
Infusion = Superficial and
Ineffective

WHO ARE WE
KIDDING...?

After 20 years, all the studies indicate that we haven't even achieved the support of most teachers for the infusion model, let alone produced the materials needed to implement it. (See the Summer, 1989 issue of "The Journal for Environmental Education.") Simply put we cannot risk devoting another 20 years to a failed approach.

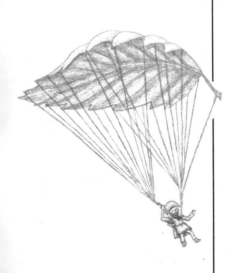

of it in math class; we worked out the lift on the wing of the plane there. And I remember we talked about the history of flying in history class too. Oh, I wrote a report on it in English. It was pretty good. I got a B." Would you stay on board? Well, folks, that's exactly what the supplemental and infusion approach is asking all of us to do. Environmental education was the most important learning of the millennium, and we were asked to do it by just sprinkling our messages throughout the curriculum and somehow the learners would put them all together and end up living more lightly on the earth. How can anyone really believe that?

You see, it is one thing to infuse some environmental messages into your math lessons or language arts exercises, but just exactly what does this mean for EE as process in the end? No one has certainly ever done this on any scale, and I doubt that anyone ever will, because we never approach any learning that we are really serious about in our societies in such a fragmented way. In practical terms, infusion appears to have meant diffusion. There's little doubt that that is exactly what has been accomplished.

And please spare everyone the sophistry contained in the line, "We teach students how to think, not what to think." Or even worse, the argument that it is unethical for us to impose our values upon them. We make them go to school don't we? We don't ask them if we are imposing on their values when we teach them how to write, do we? I want them to learn how to read. I want them to be able to describe our form of government. I want them to understand how this planet functions ecologically. And I also want them to cherish all life. I want them to use less energy and consume less material in their own lives. I want them to respect the wisdom of age, but suspect the arguments of adult comfortability. In the end, I want them to live more lightly on the earth. Influencing their values is exactly what I have in mind.

While I am at it, what is really meant by multidisciplinary anyway? What do people mean when they say environmental education is a part of all subjects? Which parts of EE are what parts of which subjects, *please?* And since about one out of every ten families moves each year (in the U.S. at least), what happens to the thousands of transfer students in this approach? Finally, it is hard enough to pin some people down on the learning outcomes for a specific subject, so what happens to measurement and evaluation when your intended outcomes are a part of everything else? How do we know if we are succeeding?

You know, it is a bit ironic when you think about it. Our EE leaders say people need to be multidisciplinary in dealing with environmental problems, but isn't that what the schools claim they were educating us to do all along? Didn't we have to take all those different courses (especially the ones we didn't like) because they would supposedly give us insights and skills from lots of different disciplines in order that we could be multidisciplinary later on in solving life's problems? I don't get it. If this approach hasn't worked, why not say so (that would be one way to free up the bloated school curriculum). If it did work, then why do it again in something called environmental education?

Somewhere along the way, when we weren't looking I suspect, someone changed the nature of multidisciplinary, from multidisciplinary in terms of problem solving to multidisciplinary in terms of learning. In other words, instead of being something you could do after you had mastered the disciplines, it became something you supposedly got enroute to mastering them. In this new version, environmental education became a sort of subliminal bonus in all your courses. (Actually, I suppose one might argue environmental *action* should be multidisciplinary, but to say environmental *education* should be would mean something entirely different).

Let me digress for a moment to explain my personal observation about what all this philosophical hot air has created. In the past fifteen years I have traveled around the world eight times and conducted almost 500 sessions on our work. I doubt that there is anyone out there who has been to more sites and centres in this field. And during that time I have learned that it doesn't do much good to ask people if they are doing environmental education. Everyone claims that they are. You have to ask them instead exactly what they do to get that job done. And do you know, when you really boil down their responses, what most people are doing that they call environmental education is conducting a couple of outside activities, putting up a poster on the bulletin board, and picking up litter in the spring. Really, that's it. All of it.

MULTI OR INTER...?

Some leaders refer to "multidisciplinary" when they are talking about the infusion approach and "interdisciplinary" when they are talking about separate EE courses that draw upon learnings from a number of related fields. In either case, you will have to examine very carefully what their learners are actually doing in order to get a handle on what is really happening. In environmental education language often masks reality.

A POINT OF CLARIFICATION...

We believe that infusing the school curriculum with an environmental perspective could serve as an important supporting objective (which is probably the most that the infusion advocates have been accomplishing all along anyway, even in the few cases where the infusion model has actually been practiced). We just disagree with the paramount status given that approach and the current benefits claimed for it. In other words, people should not confuse a general perspective or concern with specific understandings and applications (not to mention feelings). We believe many current EE leaders have it backwards. The infusion approach should be the supplemental effort, not the primary learning model.

That's what has become of environmental education in the vast majority of cases.

First of all, folks, litter pick-up has practically nothing to do with environmental education. The task is not just to pick up the litter, but to figure out where all that junk comes from to begin with in how we conduct our lives on this planet. But the major point I want to insert here is that a couple of activities, a poster in the hallway, and a litter pick-up campaign in the spring *do not* a program make.

Before I go any further I guess I should make sure you understand what we mean by a learning program. That is a much overworked term in our field. Sadly, almost any collection of activities — regardless of how insipid, unrelated, ineffectual — can be, and often is, referred to as an environmental education program. It's strange. You wouldn't claim you had a satisfactory reading program if most of your learners couldn't read at the end, or an adequate drivers training program if they ended up unable to drive, but many leaders will say they have a complete environmental education program even though most of their learners don't end up doing anything specifically to live more lightly on the earth.

We believe a genuine learning program is a carefully-crafted, focused series of sequential, cumulative learning experiences designed with specific outcomes in mind. I think those are characteristics of most learning programs, and it doesn't matter whether it is a program for learning how to play tennis, how to speak another language, how to do mathematical division . . . or how to live more lightly on the earth.

To summarize: We are convinced that supplemental and infusion has turned out to equal superficial and ineffective. By their very nature it should be clear that these methods will simply not get the job done that education must urgently accomplish. What we desperately need in our field are genuine learning programs — programs designed specifically to achieve those outcomes that environmental education set its sights on in the beginning. And we need lots of them, for different settings and situations. We simply cannot afford to wait.

"FORWARD WITH THE FUNDAMENTALS"

Does all this mean that we think there was no way to place EE in the curriculum other than as a separate course? Not exactly. For example, maybe it would have been possible for the students to learn how life functions and what that means for them in their natural science sequence, and then work on applying those understandings to crafting a new lifestyle in their social studies courses, but both these units would have had to have been highly structured to keep such content from getting lost amidst the traditional breadth of coverage in those areas.

In fact, if science education sounds like the logical place for environmental education, keep in mind that the usual scope of work there is much broader. The primary aim of science education is the process of "sciencing," i.e., gaining the skills necessary to apply the scientific method. And its purview does not usually include the task of helping learners analyze and craft more appropriate lifestyles.

Of course, scientific worksheets and apparatus offered an easy way to get the kids involved in the beginning, even if it was unclear what all their activity was actually accomplishing. One centre I heard about had the kids spend much of their visit figuring out how many leaves fell on the place in the autumn. Boy, were those learners busy. It sounds like it might have made a fairly good math activity (adding) or maybe a science exercise (sampling), but EE? Strangely, everyone seemed to buy into this. (Of course, don't forget, both the teachers and the kids got out of school to do it.) Isn't it extraordinary though how many people will accept something as significant if you just adopt the appropriate trappings?

The company that must have made out the best on this portion of the confused state of affairs in EE was the outfit that sold all those Ph test kits. Just think of the tens of thousands of those things they sold over the years. Leaders had the kids running around taking a Ph test on any available puddle. It was all great fun, but what was the point? I suspect the kids were learning the skill of doing Ph tests rather than any big picture ecological understandings. It wasn't

AT LONG LAST, NEW SUPPORT...

A recent four-year study from the Association for the Advancement of Science concludes that schools should abandon much of their memorization-based instruction in science and devote more time to helping students understand the concepts, such as the flow of energy and matter, and the interdependence of living things. (It is nice to know that the scientific community is finally speaking out.)

exactly a lifelong skill of great importance either (except perhaps for those who end up with swimming pools).

I think we are going to have to face up to it: much of science education tends to get bogged down in something other than the big picture ecological concepts. And some of our science educators turn out to be little more than technological apologists, clinging to the belief that science will yet discover the answer to all of our environmental problems. At one science teachers convention where I spoke they were handing out freeze-dried frogs to their participants as they entered the exhibit hall. What does that say about how we view the fellow passengers with whom we share this planet?

In retrospect, maybe we should have asked more often what was really wrong with adding new units or courses to the school curriculum in the first place. (Have you looked at some of the stuff they still include?) Separate environmental education units or courses, at say, the fifth, eighth and eleventh year would have probably been best. Like similar repeated material in the present curriculum, their overall objectives would have been the same, but the depth and range of their content would have increased each time. And if we couldn't make a good case back then for adding some new EE units and/or courses to the curriculum, then maybe we should have quit and let somebody else handle the job. (Of course, in the institute we would have preferred the intense, life-changing, away-from-school springboard approach that we have taken, but such experiences could have fit in easily with some regular environmental coursework.)

This does not mean either that we shouldn't have tried to infuse the remainder of the curriculum with an environmental perspective. As I mentioned earlier, it is a laudable goal but one that would have best followed the horse unlike the proverbial cart in trying to push it. In fact, a solid grounding in basic ecological understandings, along with the personal internalization for what that means in one's own life, might well provide the undergirding for other learnings. Just don't tell people the learners are going to get that by doing a couple of one-off activities from a supplemental collection.

NEW WORLD, NEW MIND...

In a compelling new book, Robert Ornstein and Paul Ehrlich propose that our environmental problems are the result of an old mind caught up in a new world, then go on to suggest how we can "evolve" a new mind for the future. There's lots here for educators to ponder, including a whole chapter called "A Curriculum About Humanity."

Besides, the point of environmental education is change; if there is no change, there is no point. Consequently, one must assume that if you were doing a proper job in your environmental education units or courses, then an environmental perspective would logically begin permeating the rest of the curriculum and the school as the learners changed how they lived there.

"lIFESTYLES bEFORE iSSUES"

We believe people can make real changes most easily in their own lives. If enough individuals are making those changes (like people did when they switched to buying smaller cars), won't the nature of the problems also begin to change? Too simplistic? I'm not so sure. Is it our task to educate everyone about the nature of the current environmental issues and all their ramifications (and in the process be multidisciplinary, I suppose), or do we educate them about how life works, and about how and why they can, and should, make changes in their own lifestyles (and deal with the environmental issues in that context)?

Perhaps we are focusing too much in our field on the overall environmental problems and not enough on individual lifestyles (particularly with the youngsters). I guess it is easier to talk about the need for listening to all the different views about land use in general than it is to talk about the problem of individual consumption in specific. Isn't it interesting though how many people quote René Dubos, "Think globally, act locally," then ask the kids to tackle the acid rain problem? For that matter, how many *adults* can absorb and utilize all the multidimensional understandings of any major environmental issue?

Sometimes it seems like the more we *examine* one of these problems the further away we get from solving it. (Do you suppose that may have become a part of someone's strategy?) After all the discussion, all the inputs, and all the sessions at a national environmental education conference on acid rain, what do you say to your learners when you return home? What do you want them to do? Understand that it's a big problem? Congratulations, they probably got the idea.

No, I am not arguing against educating people

BACK TO
THE BASICS...

Several years ago, after delivering another speech on "Sunship Earth" and its crises, I was cornered by a television reporter who asked, "Isn't this environmental education stuff really just fluff, Professor Van Matre? I mean, the whole emphasis today is on getting back to the basics in education. Isn't environmental education sort of superficial?" I replied that I was all for getting back to the basics, but I had to believe that learning about what supports our lives on this planet and what that means for us must be one of the basics. After all, is learning how to read really <u>more</u> important than learning about our place in space? Is learning how to write more basic than learning how to live within the ecological limits of the earth?

Why have we been so afraid to make our case for fundamental learnings to the parents and school boards and youth leaders out there? How about that as a theme for a major national conference?

about the big issues, but I think it is the easy way out for us. Shouldn't we focus in our conferences on developing the education programs and let other groups tackle the ramifications of acid rain? Shouldn't the issue that we focus on and call to public attention be the lack of adequate funding and support for genuine environmental education programs? What I'm trying to say is why do we have conferences on acid rain in *this* field when the average person on the street doesn't even understand the water cycle? I believe we need to help our learners build (and internalize) the big picture of how life works ecologically first, then ask them to work on their own environmental habits.

In brief, why can't we develop the broad-based environmental education programs in our own field and let the environmental action groups (that we individually support) spearhead work on the current issues? Isn't it a cop-out for us to ask a ten year old if we should ban the SST, or if abortion is the solution to population control (as one current set of materials recommends)?

As an educational consultant in the seventies, I had the task of working with a new world history teacher at a local high school. Reports were that he was spending about one-half of his course reading the Pentagon Papers with the kids. When I challenged him about the use of so much time for this purpose, he replied that most people didn't even know where Vietnam was, let alone why we were fighting there, and besides, his students were interested in the topic. I tried to get him to see that there was nothing wrong with tying into the current interest in Vietnam, but why not use what was happening in that part of the world as an illustration of some fundamental principles that could then be applied elsewhere later on. Since he had brought up the future, I added that the reason people at the moment didn't know anything about Vietnam was probably because twenty years before their world history teacher was spending all his time on the Korean War instead of focusing on the underlying principles that govern lots of similar events over time.

However, my explanations seemed to fall on deaf ears. It took me a while to figure out that this particular teacher had not learned any fundamental principles about world history himself and thus could not really understand my point. Sadly, we are so used to dealing with only the more visible manifestations of any current

"Poetry is something more philosophic and of graver import than history, since its statements are of the nature of universals, whereas those of history are singulars."

— Aristotle

issue, that we seldom take the time to examine it more deeply. Much of the issues approach to environmental education is guilty of the same offense. A group of students working on water quality, for example, often end up knowing a lot about that topic and little or nothing about other, equally important facets of their own impact upon the earth.

At one of my workshops a few years ago, a participant asked me if our approach was more accurately described as training or education. Although I didn't want to get bogged down in that philosophical quagmire at the time, I think our earth education work, like much of education in general, represents a bit of both approaches. On one hand, we want to change certain behavioral patterns, breaking some bad environmental habits and forming some good ones. And it is obvious that we are making some judgments ourselves about what particular habits need that kind of attention. However, on the other hand, we are trying to equip the students with some broad-based understandings and appreciations that will provide not only an underlying rationale for such immediate changes, but also motivate them to continue operating in a similar fashion on their own in the future.

I think the major problem we have with many of the environmental action projects that groups engage in is that the students appear to start where they should really end. Since they never developed any broad-based understandings and appreciations, nor examined their own environmental habits to see how they were contributing to the problems, these students come away with neither the education, nor the training needed for the future. Oh, in some cases, they may have used the specific issue they were involved in as a way of working back to larger principles, but for the most part, they probably ended up with some fairly narrow understandings of one particular issue. And chances are good that many of them never did make the connection between what they did at home in the evening with the issue they were addressing at school in the afternoon. Naturally, this doesn't hold true for everyone and, depending upon the teacher, some students may have become more involved in environmental concerns by coming through the back door like this. We just think a front door approach would prove more effective for more students over the long run.

WHITHER
THOU GOEST...

EE that just educates people about the environment, without asking them to make some changes in their own lives, is not EE, it's natural science.

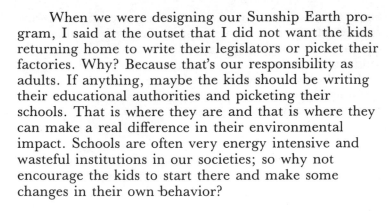

"Environmentalists make terrible neighbors but great ancestors."

— David Brower

When we were designing our Sunship Earth program, I said at the outset that I did not want the kids returning home to write their legislators or picket their factories. Why? Because that's our responsibility as adults. If anything, maybe the kids should be writing their educational authorities and picketing their schools. That is where they are and that is where they can make a real difference in their environmental impact. Schools are often very energy intensive and wasteful institutions in our societies; so why not encourage the kids to start there and make some changes in their own behavior?

Another concern we have with much of the issues-oriented environmental education is that it tends to externalize the problems. Environmental problems are viewed as the result of something or someone out there, rather than within us as individuals. It encourages the perspective that if only *they* would do this, or if *they* hadn't done that, then everything would be fine. But that's not true; we are the *they* we complain about later.

For example, one investigation project that has been written up had the kids collect samples of litter from their neighborhood streets. After analyzing it, they concluded that much of it came from the local fast-food burger place. So they put their data together and presented it to the manager of the restaurant who subsequently agreed to put out more waste containers. End of project. Two cheers for everyone.

Actually, this may have been an adequate social studies activity for how to effect change (in fact, that's what most of the environmental education issues material appears to be), but it totally misses the point as an environmental education activity. If we are really serious about the mission of environmental education, then we simply must get from litter pick-up to fast-food put down. That is the real issue, and it is certainly one where kids can make a significant difference.

Frankly, getting a group of kids fired up to make some changes in their surroundings is a fairly easy task, but when it's all over will they make any changes in their own lives? True, they may have learned a bit more in some of these projects about how to organize and implement their ideas, but what will this mean in our field if they never change their own environmental behavior? I have a feeling that a lot of these issues-oriented activities generate a spurt of short-term attention (often with an accompanying media blurb and the endorsement of some "cornucopian" group); but result in very little long-term individual change. In the end, the kids get charged up briefly about implementing some environmental improvements in their own area, while continuing to spend their money at the local fast-food joint.

Finally, I can just hear someone reading this section and saying, "Now, Steve, haven't there been some successes?" Sure there have. Although we disagree with the placement of the environmental action projects in the overall structure of a comprehensive program, it has probably been in this area where the most notable achievements have occurred. Lots of environmental projects (beyond the beautification variety) have made real dents in the environmental problems we face. And lots of teachers and leaders have labored selflessly to make such good things happen. You can find both cooperative and confrontational examples of success in schools and centres everywhere.

So I don't mean to sound like nothing good has ever happened at any centre or in any classroom, but overall it has not been enough, and I think it is the approach we have encouraged that explains why. In a sense we have squandered the goodwill and energies of a generation of teachers and leaders. If those of us who lay claim to providing some leadership in this field would have gotten our act together sooner and spoken out earlier, we might have been able to refocus the movement into more productive paths for every-one's sake.

In short, there is little doubt that every centre, and every environmentally oriented classroom has some good things going on, but due to a flawed, piecemeal approach the cumulative effect in either case seldom reaches the synergistic level necessary to promote the kind of fundamental changes needed in our societies.

"tHE OUTDOOR EDUCATION tAR bABY"

Once some folks started defining environmental education as almost anything taking place outside, it

I am afraid my comments here may make it sound like we are opposed to helping youngsters learn environmental action strategies. Not so. We support them, but we think they might best be placed within the social studies curriculum, and they might best be applied by asking the kids to make appropriate changes in their own lives instead of other people's lives. We also believe that students need to develop some basic ecological understandings and firsthand feelings for the earth and its life before they set out to effect much change in their own surroundings. We are convinced that it is this matrix of understandings and feelings out of which positive environmental action will arise, and thus rooted will prove to be most enduring.

was only natural that those involved in outdoor education would conclude that they must already be doing it. Outdoor education had been around for a long time when environmental education made its appearance on the educational scene. OE had pretty much been an extension of the schools that dealt with curriculum enrichment and application, recreational skills and outdoor pursuits, and socializing experiences. However, since these activities took place outside, and were often infused with bits and pieces of nature awareness, the outdoor education folks filled out the grant applications along with everyone else. Some of them even changed the names of their centres. Unfortunately, almost none of them changed their programs. They simply added another ingredient to their traditional potpourris.

The national activity projects in the U.S. sure didn't help clarify matters either. Even though several of them had received their funding to do something besides environmental education, they simply modified their descriptions a bit and rode the wave of interest in EE. So it was pretty easy for outdoor schools to include a few of these "fresh" activities in their programs, tack up a couple of the new posters, and add some environmental messages to their discussions with the kids. Instant EE. Practically overnight hundreds of outdoor programs and schools and centres had become part of the environmental education movement.

I hope you won't misinterpret what I'm saying. I am not opposed to outdoor education. It is important work, and I am all for it. In fact, I am for just about anything that gets kids out of school. Teachers, too. But OE is not EE. Its mission is not to help people understand how life functions here, what that means .for them personally, nor what they are going to have to do to craft new lifestyles for living more harmoniously with the ecological systems that support them.

Each year I send my group of graduate students out to visit several outdoor centres in our region. I tell them to ask the directors they meet with the simple question: "What is environmental education?" Knowing how words and actions are often two different things in this field, I also ask them to be pleasant but persistent in getting behind the words in the answer and requesting specific examples of what the centre does with the kids to accomplish what they say they are trying to achieve.

In the case of one of these visits to a highly touted nature centre in the midwest, the director finally pulled out some forms labeled "Environmental Education Workshop," saying that they had just held an in-service training session for teachers and this was an activity they shared with them. In the centre of each page appeared the title, "Environmental Education Activities — Using a Cemetery," followed by fourteen questions the leader should ask the kids about the tombstones at the local cemetery. You can probably imagine the kind of questions included: "Who are the oldest and the youngest people buried here?", "In what year did the most deaths occur and why do you think that might be?", and "How many Joneses are buried here compared to the Rodriguezes?" None of the items dealt directly with basic ecological understandings, personal connections with the systems of life, nor lifestyle decision-making. They were mostly discussion topics and it was obviously an *outdoor* education experience in the social studies.

OUTDOOR EDUCATION VS. ENVIRONMENTAL EDUCATION
(DIFFERENT SETS OF INTERESTS)

OE	EE
1. Curriculum enrichment and application	1. How do the ecological systems of the earth function?
2. Recreational skills and outdoor pursuits	2. How are we personally tied into those systems in our lives?
3. Socialization experience and group development	3. How can we make changes (individually and collectively) in order to lessen our impact upon those systems?

Again, please don't misunderstand. I am all for good social studies activities in cemeteries, but if you really wanted to, you could have a dynamite EE lesson in a cemetery as well. Think about it. It would be a great place to get in touch with how life works here.

THEMES ARE
FOR PARTIES...

One of the quickest ways to spot a problem in an outdoor program is to see if they have organized things around a series of themes or topics instead of specific learning outcomes. In many cases a centre will bring in some interns or seasonal staff, assign them some themes (like insects or soil or pioneers), then tell them to go shopping through the resources and patch together a "program" of activities. The whole thing usually takes just a couple of days to prepare, and there is seldom any specific learning outcome in mind, nor any thought given to how such an outcome might fit in with anything else. Of course, since both the learners and teachers get out of school to do this, while the activities are often led by energetic, fresh-faced college students, everyone loves the experience, but it is not enough if we are going to have a serious educational response to the environmental crisis of the earth.

"You want to see what really happens kids? Let's dig one up and look." Talk about impact! Making a connection with our own lives and the concept of decomposition would be pretty easy in that situation. And where is there a better place for getting into lifestyles and decision-making than a cemetery?

Don't get me wrong. I know we will all be in a lot trouble if you have the kids out next week digging up a cemetery, but if you're really interested in the task of genuine environmental education, then a cemetery might be a great place to start. All three components of EE could be dealt with there: how does life work, what does that mean for you, and how can you begin changing your lifestyle in order to live more lightly on the earth.

Let me reiterate: traditionally, OE has emphasized curriculum enrichment and application, recreational skills/outdoor pursuits, and socialization experiences. That's fine; those are good things to do. But it is not EE. EE was going to do something different.

When it comes to what environmental education actually said it was going to do, the main problem with most OE centres is their overall lack of clearly identified outcomes and matching learning experiences. Instead of sitting down and carefully working through what they want to accomplish with their learners, then designing and developing specific activities to achieve their objectives, they let their programs grow like topsy, adding new elements with whatever catches their fancy. Typically, such centres also latch onto the latest organizational idea, then squeeze their current potpourri through it. Learning styles, for example, swept through the field a few years ago, and everyone scrambled around to make that their new template. (Some places did the same thing with our EC-DC-IC-A formula from Sunship Earth.) But learning styles are important *after* you clearly know what you want people to learn, not before. Once again, it is a case of confusing the means with the ends.

All right, you're saying, but could you really have an EE program in an OE centre? Sure you could, but in addition to a program that focused specifically and solely on EE outcomes it would probably be necessary to model more sound environmental behavior as well. Talking about environmental education in one place,

then using styrofoam cups and paper placemats with a napkin at every seat in the dining hall just wouldn't hack it.

Just so you don't think this is an outdated problem I am complaining about, here is a synopsis of just one session at a recent conference of the North American Association for Environmental Education:

> *"Our presentation will show how . . . city schools created an education program without the expense of developing an environmental education centre. The secret was to utilize on-site locations and to adapt activities to them. The following on-site locations and corresponding activities will be discussed: cemetery — combines compass reading, orienteering, and local history; water purification plant — the drinking water cycle and water treatment; animal shelter — teaches pet responsibility and contact with animals; nature trail — tree classification, ozalid prints."*

It is pretty clear that what these folks were doing was outdoor education instead of environmental education. It is also fairly evident that they constructed their program around places and activities instead of specific learning outcomes for the participants. (Notice their use of the term adapt. That is usually a good clue about what's really happening.) In fact, it is a classic example of how most outdoor education falls short of its potential because it seldom crafts specific learning experiences to accomplish its wide-ranging and all-encompassing objectives (nor checks to see if it succeeded). Nonetheless, my major quarrel with this is in calling it environmental education and putting it on the schedule for a major environmental education conference. I just can't see what good this accomplishes in our field.

By the same token, environmental education leaders would be wise not to cozy up too closely with the outdoor education folks at their gatherings either. Although there are a lot of wonderful things going on out there, by its very nature outdoor education has something for everyone and it is quite easy to get stuck up in that milieu. Environmental education will suffer greatly if it becomes viewed as merely another booth at their educational conferences.

(In the institute we have been working on a new program for 8-9 year olds, and after almost four years we have yet to create a single activity. It has taken us that long just to sort out where we are going and to design the necessary program structure for getting there.)

If all my deprecations do not apply to your centre, that's tremendous. Please don't turn your anger on me though because they don't, but onto those places where they do. Although some of my observations here may seem picayune and overly written, I am convinced that the crucial challenge facing many centres is to either become dynamic change agents for the next century or remain social anachronisms from the last. The earth needs them like never before, but the twilight of their opportunity is not far away.

It will take a lot of courage for the staff at these centres to tackle some of the outmoded and counterproductive practices and programs they have inherited (and no doubt there will be losses), but if you are going to put your life into trying to make a difference in the health of the earth, then our outdoor centres represent one of the most vital arenas for action. Lots of places don't really walk what they talk, and you can help them see that and begin doing something about it. If you ever dreamed of being in a place and time when you could literally effect the world, then get yourself over to an outdoor centre and speak up. I really believe they remain one of our best hopes for the future.

Again, I want to make sure I am being clear. I am all for outdoor education (and its conferences), but with environmental education we are talking literally about the survival of the earth as we know it, and that is simply too important to get lost amidst all the clamorings for attention in outdoor education.

"hobby holes and stroll parks"

Okay. What happened to the promise and potential for EE in our outdoor schools and nature centres? For one thing, outdoor work of this sort has often attracted the more energetic, innovative leaders who felt too confined by the regimented structuring of mass education. Unfortunately, the same approach to education was also a great draw for those who wanted to be more creative, but less disciplined in their work. You see, any relatively unstructured field like outdoor education becomes highly attractive to people peddling their particular creative surges. Of course, that can be a boon as well as a burden. In the case of EE though, the inability of our leaders to adequately explain early on what we were trying to accomplish (plus the initial promise of government and foundation teats to latch onto for sustenance) turned environmental education into a veritable udder for those with a creative bent. For these people, camps, outdoor schools, nature centres, and similar opportunities had always represented a hobbyist's dream. They could do just about anything that struck their fancy. And the parents and teachers and directors, and of course, the kids were all overjoyed to have the benefit of any extra spurt of energy and zest for life on board.

Naturally, such people often resented any attempt to provide much structure and focus for their work, and they were quick to support the broadest possible definitions as to their purpose. Of course, if your emphasis is upon how something is learned, and you don't particularly care about what that is, then you are probably satisfied with providing primarily a recreational or social experience. Fine, but let's not mislead people. If you really want to accomplish specific learning outcomes, then you need to spell them out, identify the individual activities designed to accomplish them, and check your participants to see if they really learned what you had in mind for them. For sure, you are going to have to channel your staff's creative surges into certain predetermined paths, and you need to be upfront about this.

Sometimes you will hear people say that we stifle creativity in our work, but I don't think we stifle it, I think we channel it. You see, the difficult thing about creativity is being able to discipline yourself into focusing it in a way that produces tangible results. Some creative people never seem to succeed in doing this. They have lots of ideas, but most of them never appear to go anywhere. Their ideas ricochet off the walls endlessly, getting in the way when you least need them. In short, creativity undisciplined is like an open pit dump. It may be fun to poke around in there for a while, but it's no answer for the problems of mass education.

Besides, when people say we are stifling their creativity, what we have found they usually mean is that we are preventing them (or at least discouraging them) from using our activities in a different fashion or for other purposes. Oftentimes they've become tinkerers instead of designers. They are not really interested enough to put the time and energy and sweat into creating something of their own; they merely want to tinker around with the efforts of others and use them in different ways.

In the end, I think we are going to have to face up to another unpleasant reality too. As presently structured, a lot of outdoor centres can't seem to get it together enough to develop really solid environmental education programs. It just takes too much energy for them, or there are too many people on their staff who are used to winging it on a day-by-day sort of basis, or a few people are so caught up in their own thing that they stymie any attempt to make necessary changes. In all fairness, there are those who say they are just doing what the teachers want; that teachers don't like "canned" programs and thus believe their centre should be responding to teachers' needs. But this one quickly becomes a Catch-22 sort of argument: they sell themselves to the teachers as providing whatever the teachers want, then say they have to respond to the teachers' wants. Why can't they go out and sell the teachers on a good program that they have developed for them? Besides, what is wrong with a good

HOBBY HAVEN

DOING YOUR OWN THING...

A grad student of mine interviewed for a job at an outdoor centre recently and was told that they wanted to create their own program there. The director said they thought they were fairly innovative themselves and they wanted to do their own thing. However, as it turned out, they were only going to give their new staff member a couple of weeks at most to put a "program" together (and a minuscule amount of money to support it). Of course, there's nothing wrong with wanting to build your own program like this (as you'll see, we are encouraging people to do that in this book), but why can't we get such places to face up to what it takes to really do it? Program building is tough, demanding work, and it requires considerable time and resource.

MY APOLOGIES...

This section is going to upset a lot of people, but I don't see any way around it. You see, I have mixed emotions about many of our outdoor centres. On one hand, I know that the leaders in these places have often worked long and hard against amazing odds to keep things happening at their sites. On the other hand, I am often appalled at how much traditional victorian nature study still permeates their entire operation, even when they talk a different line. Somehow we have to get these places to take a closer look at what they are really doing, before it's too late.

I also want to make sure that my comments don't reflect unjustly on the staff at these sites. In my experience these people are almost always environmentally concerned and committed, but they have often either let themselves be misled by supplementalist thinking, or rationalized their personal interests as having significant educational merit, or become subjugated by the politicized agencies or boards that control their destiny. Fortunately, there appears to be a new generation of potential leaders in many of these places. If we care about the

canned program? Good teachers used canned programs every day. If you were really serious about learning how to play tennis, for example (or helping someone learn how to read), would you go to people who had spent much of their life working on coming up with a good canned program to do the job, or would you really prefer the temporary leaders who put together a few ideas last week?

Besides, who are we really kidding anyway with that tired old line about responding to what the teachers want? Crafting good educational experiences takes a lot of time and energy. So let's quit misleading the teachers by claiming that we can come up with good experiences for anything they want. Most importantly, let's quit being dishonest with ourselves.

Finally, please don't tell me, "Well, regardless of what you say, Van Matre, the kids like it." I don't think that argument holds up either. Of course, they like it. They're not in school! You could strip them naked, pass out sticks and turn them loose in the forest, and guess what, they would return home saying, "Ya-ha, did we have a great time." My least favorite argument though is the one about how the teachers like it. C'mon folks, they're not in school either. Here is a fundamental: anytime the kids (or teachers) get out of school to participate in your program, you have everything going for you. It would be astonishing if they didn't like it.

I am continually amazed at how many of our large outdoor centres claim they are doing environmental education when in reality they are offering little more than a field trip service that depends heavily upon the natural wonders of their site. In the states, this is particularly true in the national parks. Don't get me wrong. There is nothing inherently bad about taking kids out for a hike to one natural phenomena or another (in this case it's how you go about doing it that counts), but palming that off on environmentally concerned folks as environmental education is a crime. It is just not adequate. The tragedy is that many of these field trip services, regardless of what they are called, often have large budgets, sizeable staffs, and considerable public goodwill, but they could, and should, be doing so much more. In reality their offerings are little more than the field trips of yesteryear now served up in a fancier setting. I am sure the kids get some wonderful

natural hits this way, and that certainly has to have some value, but if we could get these places to engage in some genuine programming (not supplemental spice), including built-in behavioral changes back at school and home, then such places could really set the standard for the rest of our field.

Truthfully, I don't know what all this means for the future. Our outdoor schools offer unparalleled opportunities to make a real difference in their communities, but until many of them either gain new leadership or reconsider their approach, their prospects for taking an active and decisive role in the original mission of environmental education appear rather bleak.

earth, then we must single out these folks and encourage them to rise up and change the course of the current educational response to the environmental crisis of the planet.

"Today we'll hike to the waterfall and talk about ..."

As for our nature centres, too many of them have become little more than stroll parks for the more affluent. As soon as you enter, you can sense that these places are not really interested in changing people's lifestyles. Their primary goal seems to be to provide a sanctuary for those who like to put on their tweeds and binoculars for a Sunday afternoon nature walk.

Their staff is often made up of paraprofessional naturalists who want it both ways, that is, they want the status of being a scientist (although they don't really practice much "sciencing"), and they want the jobs of educators (although they don't really draw much upon

NATURE CENTRES SUCCUMB...

Once again, I am sorry, but asking an association of museums to accredit them is symptomatic of the malaise already infecting many of our nature centres. We must purge ourselves of the continued victorian desire to conquer the world by identifying and collecting and exhibiting pieces of it.

You see, at heart a museum is a repository, a collection of things to experience, but a nature centre should be a jumping off place, a springboard for experiencing natural <u>communities</u>, not collections. Please don't misunderstand: museums are important places. The good ones are like time machines able to whisk you artificially through space, but nature centres should be rooted primarily in the here and now, immersed in the natural flow of life.

You can gussy a museum up in a lot of ways (and I am writing about that in my next book), but a museum is a warehouse of sorts. It encloses things for you, while a nature centre is supposed to help you open them up, to experience them in a larger, natural context. In The Institute for Earth Education we plan to begin accrediting nature centres based upon their ability to turn people on to natural systems and communities through firsthand experience, their skill in conveying the big picture of how those processes and places function, and their success in helping people lessen their personal impact upon them.

what we know about good learning). You will probably find that many of these centres still offer primarily identification experiences (focusing on the pieces of life instead of its processes), an arts and crafts child-minding service for local parents, some assorted outdoor education activities — often based on relatively meaningless and boring worksheets — or maybe even a couple of unrelated environmental games, all topped off with a few imprisoned, neurotic animals, and backed up by an array of environmentally questionable fund-raising activities. In all probability, a gift shop will also dominate their visitors centre, and other than a couple of posters nothing much will really challenge people to change how they live on the earth. Welcome to the most popular nature centre in the Age of Excess. It may be just down the road.

Don't take my word for this. Check one of these centres out for yourself. You can almost always spot the personal hobbies of its staff and director; such things as collecting or training animals, creating mini-habitats, teaching college science courses, restoring a prairie, etc., will likely claim large chunks of space and time. And all of these pursuits, although worthwhile ends in themselves perhaps, will probably be siphoning off much-needed energy for implementing any focused educational programming.

If all this is too abrasive, I'm sorry. I know it is not a fair characterization of many nature centres. (It wasn't meant to be). And I am sure every centre has some good things happening; things they can be justifiably proud of offering. But it is just not enough. We are in trouble out there. Nature centres should represent one of our primary sources of genuine environmental education, but many of them just don't make the grade. (To give you an idea of how some actually view themselves, the American Association of Museums has begun accrediting them in this country). In short I don't think the environmental movement can rely on many of these places for the kind of focused, sustained effort that will be necessary in the years ahead. We had better start thinking about creating some new kinds of centres in our field, or radically altering the present ones.

MUSK TURTLE
Sternotherus odoratus

EVALUATING AN OUTDOOR CENTRE

☑

☐ Directors and staff clearly model environmentally-sound practices and lifestyles (or there is evidence that they are working on them).

☐ Programs are designed with specific outcomes in mind and use stimulating educational techniques and tools to pull instead of push the learners.

☐ Everything works together and the visitors and participants are oriented to the sequential and cumulative ways in which they can proceed.

☐ Programs and exhibits focus on major ecological concepts, such as, energy flow and cycling, and connect these processes to the daily lives of the participants and visitors.

☐ Facilities are environmentally-sound in design and operation (or are being retrofitted to become more so).

☐ Programs and exhibits challenge participants and visitors to make changes in their own lives and model possible choices for them to consider.

☐ Overall atmosphere conveys great care and concern about the earth's places and processes, while promoting a sense of wonder and adventure for natural areas (emphasizing magic and meaning instead of names and numbers).

☐ Programs serve as carefully crafted and focused "springboard" experiences (in ecological understanding and feeling) that must be completed later back at home and school.

☐ Regular workshops or courses are offered in crafting more harmonious lifestyles (organic gardening, environmentally-responsible investing, modifying homes, etc.).

☐ The characteristics described above dominate the overall feel of the place (from entrance to exit) and there are few, if any, discordant notes (unnecessarily caged animals, environmentally-unsound materials and practices, racks of trinkets for sale, etc.).

Each item is worth ten points. Rate your local centre and see if it makes the grade. Be sure to share your results.

Shiela Harty, at the Centre for the Study of Responsive Law, pointed out in this revealing book the extent of the environmental propaganda problem in our schools. (Her description of the origins of Project Learning Tree is of particular interest.) Unfortunately, she failed to recognize how far the "environmental educators" working on some of those industry-sponsored materials had already succumbed themselves to cornucopian messages.

WHO'S PAYING
OUR BILLS...?

In the U.S. timber industries fund Project Learning Tree, electric power companies provide the support for the Association for Nature Centre Administrators, game commissions back Project Wild, etc. We're in a lot of trouble out there.

"CO-OPTED by the CORNUCOPIANS"

In all honesty, I also fear that many of our professional positions and organizations have already been infiltrated by the view that the earth is our horn of plenty and all we will have to do is a better job of managing it. Granted, infiltrated is probably too strong a term, but the names of the companies and bureaucracies that a few years ago would have raised more than an eyebrow in our field often appear prominently now on the credit pages of our materials, the exhibits at our conferences, and below the names of some of our most visible leaders. If we are serious about our mission, we had better start asking ourselves why some of the agencies and industries that helped create our environmental problems in the first place, are now suddenly sponsoring things in the environmental education field, and what that means for our future.

Project Wild, for example, is a collection of supplemental activities that has come under increasing fire from both animal welfare and animal rights groups, and has been strongly criticized for its subtle emphasis upon management as the only viable approach to our relationship with the other life on this planet. (As a Canadian observer put it, Project Wild should really be called Project Tame — *Towards A Managed Environment*.) Yet many environmental education organizations appear almost as if they are being held hostage by the funding available in that network:

⊕ *They give over chunks of their newsletters to Project Wild events.*

⊕ *They block out entire sections on their conference schedules for Project Wild activities.*

⊕ *They end up with practically interchangeable leadership between Project Wild and their own supposedly independent professional organizations.*

You see, there were lots of people out there in the seventies who aspired to become environmental education leaders, but they were often left floundering around with nothing much to do. Then almost overnight many of them became "trainers." They were invited to special sessions and given free materials to dispense. They had support money available for attending conferences and conducting workshops. Most importantly, they had learned at long last how to tap into agency and industry budgets (keep in mind that for the environmental education stepchild this was something akin to finding the Holy Grail in a box of hand-me-downs).

I know, all this sounds rather brutal, but I don't mean it that way. These people were good people (I hope I can still count some of them as friends), and most of them no doubt had their hearts in the right place, but I think they let themselves be misled about the real nature of the "projects." They sold out without knowing it. Only the most honest and courageous leaders out there will be able to confront themselves and challenge what has become of their "professional" roles and organizations.

BE PREPARED...

Both The Institute for Earth Education and the American Humane Society have produced lengthy critiques of Project Wild. Reprints are available from our respective international offices. In addition, some Project Wild folks may lead you to believe that they have revised their material thus satisfying their critics. This is simply not true.

tAKING a STAND
(WHAT YOU CAN do)

⊕ Join your regional and national associations and speak out at their events about what needs to be done (attacking the problems, not the people). Examine their newsletters to see what they inform their members about (and what they leave out).

⊕ Write to your educational authorities and object to random, superficial responses to our environmental problems (offering to serve on an advisory group).

⊕ Examine educational materials looking for subtle management messages and "cornucopian" solutions (explaining to the authors how they can solve the problem).

⊕ Notify your elected representatives that you oppose spending public funds on ineffective supplemental collections of environmental activities, and the offices to distribute them (letting them know there is an alternative).

⊕ Ask those passing out educational materials who is paying for them and why (reminding them that there is no such thing as a free lunch).

It would be different too if the project materials were passed off for what they actually represent, i.e., supplemental collections from which most teachers only use at best a couple of activities. But no, these collections are held up as if they are the answer for EE. State coordinators devise elaborate curriculum plans based on the infusion model, then push the projects as if they were actually designed to accomplish those aims, and finally claim that they have EE in their state when the vast majority of their teachers still do practically nothing. It is a serious educational fraud that's being perpetuated in many places. And it is costing the taxpayers a lot of money. Somebody should call them on this.

Regardless of any disclaimers I might add, some will read these words as merely the whine of sour grapes, but I would hope that our indignation might be seen as a heartfelt challenge to the field instead. Believe me, we take no pleasure in setting ourselves up as a lightning rod like this. We have little to gain and much to lose, for we realize we are not perfect ourselves. But in all good conscience we can remain silent no longer. If we are serious about the original mission of environmental education, then all of us have to get farther away from pushing management messages and closer to modeling personal choices.

"THE MANY DISGUISES OF DR. YES"

Lots of new verbal camouflage has begun appearing in the past year or so that either alters the real nature of what the supplemental approach has to offer, or asks you to accept that approach for reasons other than what it actually represents, or cleverly hides the truth about what is really going on. Here are a few of the pretenses that you should be on the lookout for:

⊕ One of the most misleading assertions to appear on the environmental education scene has been the position that "We respect the professionalism of teachers" (by providing them with the activities and then letting them decide how and where to best use them with their learners).

However, when you stop and think about this statement, does that mean that all those reading and math and science programs did not respect the teachers? Of course not. It means they realized that many teachers just don't have the time, nor the resources (and sometimes the skills), to develop complete programs, and thus would appreciate being freed up from that chore so they could focus on implementing them instead. (Being a professional teacher doesn't mean that you have to design and develop all of your own learning programs.)

If we really want to treat teachers like professionals, then we should give them something to work with — at least a foundation and a structure upon which to build — instead of an assortment of activities and the direction to "do anything you want with this stuff." That's not respecting teachers; it is condemning them to undertake a building job without any tools or instructions — not even a picture of what the result should look like — just a pile of materials to sort through.

In reality, we should all pity the poor teachers who get treated like professionals in this way. In the supplemental collections of activities they receive, you will usually find no directions for building complete programs, no models for what such programs might look like, no instructions that there are some elements that may be more important than others, and no suggestions for how to achieve particular behavioral changes in the end. (Strangely, this doesn't prevent folks from defending their activity packages by claiming that, "We're all trying to reach the same goal.")

⊕ *"We believe in diversity."* Watch out for this one. Some leaders will pull out this argument to support themselves when all else fails. The implication is that since we support ecological diversity, we should naturally support diversity among programs. First of all, this is an apples and oranges comparison. We do not believe you can classify a collection of supplemental activities as a learning program to begin with, so we don't think it is fair to

ask us to support it in the name of diversity. Second, what we are actually talking about here are educational products, and even if they all were genuine learning programs, it does not follow that they would all be good ones, nor that they would all merit the expenditure of public funds. Keep in mind that many of the environmental education products available today are not dependent upon the marketplace of ideas for their success. They are heavily subsidized and promoted by various cornucopian agencies and industries.

WHY NOTHING MAY BE BETTER THAN THE "PROJECTS"

It has become popular to defend Project Learning Tree, Project WILD, etc. in some quarters by saying that they are better than nothing, but that may not be the case. Here are five reasons why the reverse may be true:

1. *It depends on your activities.* The problem with a blanket statement like this is that it tends to relieve people of any responsibility for judging the merits of the activities to be used. For example, if the activities selected will subtly convey the idea that the earth is our cornucopia and all we have to do is a better job of managing it, than our learners would probably be better off without that message. After all, they already get enough of that viewpoint from most contemporary advertising without including more of it under something called environmental education.

2. *This kind of statement assumes that without the activities referred to there wouldn't be anything, but that doesn't necessarily follow.* In fact, there might be something better. Sometimes good things in this field get overwhelmed by all the hoopla paid for by the cornucopian agencies and industries. Besides, since many of the PLT or WILD activities are based on fairly common ideas, if left alone, lots of teachers would probably come up with them anyway, and if given different guidance, they might actually come up with something far more focused and effective.

3. *When the materials being compared serve to reinforce a dysfunctional perception of the field, one that may end up contributing to our environmental problems instead of solving them.* Since most teachers who have been introduced to the "projects" will use only a couple of the activities in those collections, it reinforces the idea for

them that environmental education means just including a couple of one-off activities in their lesson plans. Sadly that perception may actually cause more damage than any good achieved by the activities themselves.

It would be different if the projects started out with a section on building sequential, cumulative learning programs with their activities (along with an explanation of why doing it is so important), but that's not the case. Instead teachers are encouraged to just pick something that strikes their fancy and do whatever they want with it. In the end, it's like claiming you have music education in your school because you sing a couple of songs with the kids in the autumn and play a game about Beethoven with them in the spring.

4. *With "nothing" most leaders might at least remain open to something else.* Since the value of a couple of isolated activities is so marginal (even if they were the best of the lot), it would probably be better for the teachers not to do anything if that meant they would be receptive to a serious educational response to our environmental crisis when one appeared. Unfortunately, by claiming to do the job of environmental education, supplemental collections of activities often drain away any motivation that would lead teachers to look for authentic alternatives. After all, in their eyes, they are already doing the job. And since our leaders have led them to believe that environmental education is, by definition, just an assortment of such things, they will probably not look very closely at other options that come along.

(In the same vein, the existence of the projects also lets our departments of education and conservation off the hook as well. Many of them support those supplemental efforts and thus feel they have nothing more to do.)

5. Sometimes the alternative can be so ludicrous as to make the comparison with nothing sound like little more than a hollow platitude. For example, if your house is on fire and your neighbor shows up with a cup or two of water to throw on it, you probably won't say, "Thanks, it's better than nothing." Given the obvious need, a couple of cups of water would seem so ridiculous that you would probably tell your neighbor not to bother, especially when it would be so easy for someone to do more. In fact, you would probably tell your neighbor, in no uncertain terms, that you would appreciate and expect more assistance than that. The same holds true for an educational response to the environmental crisis of the earth — a couple of activities are so minimal as to be almost meaningless. We must not allow people to get away with using this argument to absolve themselves from asking more from their neighbors . . . (in no uncertain terms).

A ROSE IS NOT A ROSE...

Be especially cautious in our field whenever you hear a supporting argument that uses an ecological term or phrase. It may sound good, but you should ask yourself if it's really applicable to the position being defended. Besides, cornucopian thinking has already latched onto most of the good words available.

One variation on this position that you may run across is the contention that goes something like this: *"We're all in this together. We all have the same goals. And because of the fragile status of environmental education, we should all be supporting one another."* Wouldn't you love to have a bank note for every time that argument has been put forward historically to stifle dissent? If you really believe one approach to solving a problem is wrong, if you really believe it misleads people about the true nature of the problem, if you really believe that it may actually exacerbate the problem instead of solving it, then don't you have some obligation to speak out about your convictions?

Another version of this one that you will hear from those who would have you say yes to everything is summed up in the statement: *"We shouldn't be so negative."* And why not? I can think of a lot of situations in the past where progress was made only after someone was willing to stand up and say enough is enough. No more. This is wrong.

One leader even wrote to me saying that he will only allow his students to take positive positions (I trust he is not one of those folks who also claim we should not impose on their values.) Do you suppose he just ignores the lessons of thousands of years of recorded history, or does he ask his students to ignore them? Positive positions are great, but sometimes individuals have to be willing to put themselves on the line — to take the negative stance as well.

In the same vein, I am sure you have heard someone take this approach: *"We shouldn't get bogged down in petty differences."* Hey, these are not minor points we are talking about here. They go to the very heart of what we are doing, and mistakes now will cost the earth dearly in the future. In the United States the infusion advocates have created supplemental collections of activities and called them programs. Now they are creating a network of national centres to distribute these counterfeit works. The point is our differences are not insignificant.

Here's a different kind of disguise: *"We can't do that stuff here because it has too much impact on the environment."* That's what they told my grad students at one of the centres they visited. But when they pinned the staff down on exactly what they did with a school group, it turned out that three of their five activities were ours masquerading under other names.

Please don't misunderstand. I don't mean to belittle the concern about the possible environmental impact of our activities. It is an important consideration. However, it has been our experience that most places can deal with that question without sacrificing the implementation of a genuine program. There are always adjustments that can be made to an activity that will help minimize the problem.

For example, I recall the staff at one national park telling me there was just no way they could take people off crawling around on the ground to set up our Micro-Trails activity because the area was too fragile. They said they had no choice but to keep everyone on the asphalt path. I suggested that in this situation they should try getting kneepads for their participants and letting them lay out their micro trails as they crawled along on the asphalt, reaching out as far as they could to set up the station markers for their miniature trails. This way they could minimize their impact without sacrificing the close-up observation and contact that the activity provided.

Nonetheless, regardless of what you do, we know there will still be an impact. It's always a question of trade-offs. Did the learning outcome you achieved warrant the impact it necessitated? The leaders at each centre will have to answer that question for their own setting and situation. The important thing is that we ask it instead of using it as a

WATCH OUT...

One of the most worrisome developments currently underway on the U.S. environmental education scene is the attempt to create a network of national EE centres. Not only will such a development likely perpetuate much of the current nonsense, it will institutionalize even further an approach that has little hope of solving our environmental problems. Once embedded in the "cornucopian" bureaucracy, where it can be supported by an endless stream of agency and industry funds, the supplemental and infusion approach will have little need to justify itself. Consequently, there's little hope that it would be recognized for the failed mission that it actually represents. As the old saying goes, "If an idea is bad enough, it will stay around forever." In this case, we're likely to assure it.

crutch, and then work to come up with the best possible solution whenever there is a problem.

⊕ Don't be mislead either by the *"Where's your research?"* question. Not surprisingly, given the costs involved, long term evaluation is practically nonexistent in this field, but doesn't it make sense, at least, that you would be more likely to get such behavioral change if that is what you were aiming for in the first place? Some leaders are fond of asking about your data, but surely you will have a better chance of producing people who will live more lightly on the earth if you tell them that's your goal and begin helping them do it. Besides, what is the option? Not to try for long term change, or wait until you can afford a costly study before getting started?

There have been three short-term studies done on our Sunship Earth program, for example, but we are still unsure what happens to those learners five to ten years later. Nonetheless, we feel confident that we are closer to our goal than if we had never even asked them to begin living more lightly on the earth. Frankly, the supplementalists may use some great holistic phrases in their introductions, but the fact is, their actual learning experiences seldom ask anyone to make some specific changes in their lives.

⊕ Finally, there is the biggest cover-up of them all: *"We aim to be a part of the mainstream"* of public education. The underlying implication here is that you can't get genuine programs instituted in most schools, or you can't take the kids away from the buildings, or you can't get the teachers out of the classrooms, so you must design supplemental activities that don't require much planning or effort. Folks, this self-fulfilling prophecy has done a lot of damage in environmental education. You can get genuine programs implemented; you *can* take the kids away from the buildings; you can get teachers out of their classrooms. (In fact, most teachers are just as anxious as the kids are to get out and get involved in a good educational experience.) And we have lots of evidence to support that this is true. You just have to decide that that is what's important.

Besides, being in the mainstream today is more likely to be a part of our problem rather than its solution. If all the vast sums of money that have been spent on supplemental materials over the last twenty years would have gone instead to modeling and supporting some genuine programs, we could have probably already turned the tide in the environmental movement. I doubt that the future will look kindly upon us because of our pusillanimous vacillation and deception on this point.

"ACCLIMATIZATION ACRIMONY"

It would be unfair of me to lash out at all the problems like this and not comment upon our own shortcomings. We have certainly made our share of mistakes. Here are nine of them:

First, we are well aware that we must bear some responsibility for the present state of affairs in the field, for although we have consistently sought the ends described here, we have not always done so in a clear-cut programmatic fashion. All too often our activities have contributed to the morass instead of helping make sense of it.

We always viewed Acclimatizing, for example, as an extension of the original six hour program introduced in Acclimatization. However, many folks saw it as merely another collection of supplemental activities. (Like its offspring, that probably explains why it always sold so well.)

Second, there are far too many examples of playing "twenty questions" in our original materials. (We'll look at that problem in some detail in chapter five.) I just failed to grasp what would happen with that technique in the hands of less focused and enthusiastic leaders (or more didactic and manipulative ones). I look at

"The reasonable man adapts himself to the world; the unreasonable one persists to adapt the world to himself. Therefore, all progress depends on the unreasonable man."

— George Bernard Shaw

some of those descriptions now in <u>Acclimatization</u> with an embarrassed grimace.

Third, our initial school program, Sunship Earth, was too long and too leader intensive for most schools to adopt. Unfortunately, it was also released just as the interest in long-term residential experiences was declining, and it took us too long to adjust to this reality and show people how they could use some of the parts of Sunship Earth in building their own programs without the necessity of offering the whole. (Our most recent programs, Earth Caretakers and Earthkeepers are designed to address these problems. They require just one or three days away from school respectively to initiate and can be conducted by a couple of leaders.)

Fourth, although we engaged in the kind of evaluative research that anyone would employ in developing an educational program, we did not spend enough effort documenting our results formally. We did urge everyone who set up a Sunship Study Station to also evaluate the program for their own learners, in their own setting, but in retrospect, we should have done more to support this and to share the outcomes in the usual journals and conferences.

Fifth, many of the Sunship Earth programs established simply did not get enough transfer back to the school and home settings. In some situations they turned out to be just another week away from school that still had marginal relevance for what occurred afterwards. (That is why our ongoing development team has gone back and re-worked Sunship Earth so it cannot be completed at a site. The action must continue back at school.)

Sixth, for many years we maintained a closed, limited membership organization in the institute. This kept out lots of good people who would have contributed much to our growth, both internally and externally. As a volunteer organization with no grants nor government support, we were reluctant to get others involved faster than we thought we could handle them. We should have risked more.

Seventh, perhaps our major problem, though, over the years has been in trying to provide all the programs ourselves. In short, we tried to do too much

with too little. Today, we are more aware of our own limits. We have to put greater emphasis upon the structural template of earth education and work harder at helping others use our program-building tools in developing their own programs based upon that structure.

Eighth, we also spent too much time in the early years dealing with various bureaucratic functionaries (like the national Youth Conservation Corps staff) who didn't really understand nor care about the long term mission of genuine environmental education. I am afraid we were guilty of chasing the money like everyone else.

For example, in 1979 we submitted a lengthy grant application asking the U.S. Office of Environmental Education to provide funds to aid us in disseminating the Sunship Earth program. Please note that we didn't ask for monies to develop the program; we had already done that part of the job ourselves over several years and at considerable sacrifice. We merely asked for financial help in letting others know more about it.

The result? They turned us down, saying Sunship Earth was *not* environmental education. Although our disenchantment with the field predated that decision, I can see why some may feel our bitterness at this outcome sharpened our attack upon what had happened to environmental education. (A review of the U.S. Office of Environmental Education tragedy lies beyond the scope of this book, but the whole effort appears to have been a singularly unproductive, perhaps counterproductive, chapter in the lost mission of EE. Luckily, it was eliminated.)

Ninth, we spent too much time worrying about how some people were using our activities and not enough time coming up with the models for them to use. (I must admit this one remains a problem. We still cringe when we see our activities pop up somewhere out of context, but we're doing better.)

Okay. Do we think we have all the answers now? Of course not. We don't even know all the questions yet. However, we have decided to start over. We are going back to try to achieve what environmental education said it was going to do and didn't. We have changed our name, broadened our base, and invited like-minded folks to join us. We are calling our work earth education.

Do we think our way is the only way to get there? Of course not. But we do think that paying attention to exactly where we are going is a big advantage, and we remain openly suspicious of those who claim they are aiming for the same goal when there appears to be no specificity in where they are going, nor any consistency or continuity in their methods for getting there.

In our work, we have elected to design and disseminate "springboard" programs, i.e., highly-charged, focused educational experiences that serve as a springboard for what will take place for the learners back at school and home over a period of several weeks. As you will see in later chapters, we also feel strongly that we need to get the learners out there and "immerse" them in the sights and sounds and smells of our richly textured natural systems and communities.

However, we do not believe our programs are the only answer. For just a fraction of the typical costs of a supplemental collection, someone could design a dynamite school-based program. Of course, we would still argue that you have to get the learners out there in touch with the systems of life, one on one. Trying to convey to a kid in an urban classroom what life is really like out there would be like trying to explain life on the earth to a termite without ever getting it outside the termite mound. You just can't *feel* your true relationship to the systems and communities of life here without immersing yourself in them. But at least a school-based *program* would be a step in the right direction for "getting there."

I hope it is clear by now. We believe environmental education became a bit of everything to everyone, and consequently not much of anything to anyone. It became anything anybody wanted it to be at the moment and nobody said there was anything wrong with that; instead, everyone collected their grants and awards and fees and moved on. Yes, that is pretty rough, and of course, none of it was done maliciously (although I suspect carelessly), and perhaps some of it was merely the unavoidable settling out that must occur in any new field. It may have also been inevitable that in the midst of the euphoria generated by a new movement with its accompanying new positions and funds, that its leaders would be reluctant to blow the whistle and call for a reassessment of their purpose and direction. However, the dream was so important, the mission so

urgent, that we should all probably be just a little bitter about the lack of hard thinking and public declaiming evidenced by many of our contemporaries, and angry at ourselves for not speaking out earlier as well.

SEVEN rEASONS WHY ENVIRONMENTAL EDUCATION fAILED

⊕ Defined its objective so broadly that almost anything could fit somewhere within it;

⊕ Promoted a supplemental, infusion approach instead of genuine, focused educational programs;

⊕ Encouraged short term projects based on the issues, while largely ignoring the long-term lifestyle decisions of its learners;

⊕ Accepted the funds and sponsorships available from the "cornucopian" agencies and industries that helped create the problems in the first place;

⊕ Neglected to clearly distinguish itself from other groups interested in outdoor experiences;

⊕ Provided no guidance for why some ecological concepts may be more important to convey than others;

⊕ Generated mounds of conference paper, but no clear vision of a model to aspire to.

Please don't misunderstand: it would be different if I thought all of this confusion was merely the product of some sort of educational evolution in an emerging field struggling to define itself. In fact, the harshness of my words is probably in direct proportion to the lack of any such struggle at the practitioner level in environmental education. As we see it, precious little assistance has been offered in helping the people actually working out there in the field sort out what they are doing.

So earth education is not environmental education. It is an alternative to it.

E.E. IN THE BUILT ENVIRONMENT

What does all this mean for you? If nothing else, I hope we have given you some new insights in this chapter about the original mission of environmental education and its present status. As Albert Einstein put it, you don't have anything to think about unless you know there is a problem. Believe me, there's a problem. Won't you join us at least in thinking about it?

CHAPTER two

aCCLIMATIZATION...
a
SENSE
OF
rELATIONSHIP
WITH
tHE
eARTH

For many years I began our workshops on what we were doing in nature education by going to the chalkboard and writing the words

ON'S → TO'S

and saying, *"I'd like to begin our session with a phrase you'll hear us use a lot in our work: what are your On's and what are your To's?"*

"In other words, when you are designing an educational experience, what are you building on — where are your learners coming from — what kind of backgrounds do they have — what understandings, skills, appreciations have they already developed? What are you building <u>on</u> with a particular group of learners and what are you building <u>to</u>? What do you really want them to have when they are finished? And can you spell that out with the same kind of specificity, saying this is what you want them to have at the end — this particular understanding or skill or appreciation? As I used to say to my staff, 'What do you want them to have to put in their pockets and take home with them at the end of this?'"

"We think it is pretty important to sort this out in mass education (whether it's for a single activity, an entire day, or a whole program). What are you building on and what are you building to. . . ."

"In a sense, I am in trouble at this workshop because I'm not sure of what I'm building on. I think the best thing I can do to get us started is to give you an idea of where we're coming from with what we call Acclimatization. So this opening session is designed to build a foundation for the rest of our time together. What is Acclimatization; where are we coming from?"

The same rationale holds true for this book. Now that I have made our case for why we believe environmental education went astray, it is time that I give you an idea about where we are coming from ourselves.

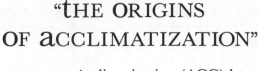

"tHE ORIGINS OF aCCLIMATIZATION"

As a program, Acclimatization (ACC) began in the northwoods of Wisconsin in the early sixties. I arrived on the scene at a private boys' camp there after my sophomore year in college, and the magic of those lakes and forests under a summer sun captivated me immediately. Although it was not my first counseling job, there was something about the land and life at Towering Pines that spoke to my deepest feelings and set the stage for events that would have totally unexpected results.

Among other tasks that summer, the owners asked me to take on setting up a new nature program. When I asked what they had been doing with nature study in the past, they hesitated a bit and finally said they really hadn't done much of anything for the last couple of years. "We used to do some things — the counselors took the kids out and identified things, collected things, dissected things, experimented with things — but to be honest, the kids just didn't seem to care very much about all that." (At this point, perhaps I should digress to explain that the camp was made up of over a hundred boys ranging in age from about seven to seventeen, and they were there for about seven weeks.) So I said I would see what I could do, and thus set out that summer to lead a small group of kids in some outside experiences.

Since one tends to do what one has seen done, you can probably imagine what I did with the kids in the beginning. I took them out and identified things, collected things, dissected things, experimented with things, and guess what, the owners were right, the kids really didn't care about all that stuff. Oh, you could get three or four to show up — five or six on the days you cut something up — but for the most part, the kids weren't much interested. They were more attracted to sailing and tennis and riding and water skiing, and so forth.

Finally, somewhat out of desperation perhaps, I sat down in a corner one day and asked myself a simple question, what are we building *to*? What do we want these kids to take home with them at the end of the summer? You see, I had a pretty good idea by then about what I was building *on*, but what was I really building *to*? As it turned out, my answer changed everything we would ever do. I decided that what I really wanted to do was just to turn the kids on to the natural world. I wanted to convey my love of the earth and its life, not for its labels and fables and fears, but because of my rich firsthand experiences with it. I wanted to convey a feeling of at-homeness with the earth, a feeling similar to what you have in your own house. You know what I mean, you feel good there; you understand its moods, its smells, its nooks and crannies. You return each evening, open the door, and say, "Hey, home." I wanted the kids to have that same feeling of security and comfortability that they have in their own homes, but with the planet itself — our preeminent home — the earth and its communities of life.

In fact, that's where I got the term Acclimatization. I decided to call the program by that name because at the camp we were trying to acclimatize the kids to the earth and its natural systems. To acclimatize means to become accustomed to a new setting or surrounding, and that's what we were doing. For many of our urban-suburban youngsters the natural world was a new place, a place they had never really experienced very deeply.

Consequently, we organized our original ACC program around the natural communities of life that surrounded us. You see, when I looked around camp — and since then I've noticed the same thing at hundreds of other camps and centres around the world — it appeared to me that someone had come in, bulldozed a clearing out of the relatively natural communities there, and built a camp in the middle of the "hole" they had created. As a result, most of the kids spent most of their time in the buildings and on the playing fields in that hole. They had very little actual contact with the natural areas of life all around them. So I decided to introduce the kids to these natural communities they were missing. In the beginning then, Acclimatization was based upon several hour-long, introductory experiences in the marsh, forest, lake and bog communities of our northwoods setting.

(Please note that our natural areas were large enough that we could take the groups to different places in these natural communities and thus minimize our

Acclimatization documents our original barrier-breaking introductory nature experiences for youngsters and the pioneering methodology on which they were based. First published in the late sixties, it became a landmark work in the field.

impact. Please keep in mind too that this was a boys' camp, and the characters in most of my initial stories will reflect this fact.)

Right away though, we discovered there were some barriers between our on's and to's; some barriers that the kids came with that we would have to overcome. The first one was an attitudinal barrier, a built-in disposition towards the natural world that we would have to deal with before we could do anything else. Let me tell you a story to illustrate the point. I was up in a tree one day watching a group of kids when. . . . Maybe I had better back up for a moment and explain first why I was up in a tree. Unfortunately, every time I took the kids into the marsh myself, I would get so wound up in the doing that afterwards I wasn't really sure what had happened to them. I didn't have any perspective on the learning situation. So one day I got hold of an energetic counselor-in-training who was interested in the natural world, and said, "Jimmy, you take the kids out there and try out some of these things. I'm going to crawl up in a tree and watch."

Folks, I strongly recommend that to you. If you really want to see what goes on in a lot of outdoor learning situations, just pull back, hide somewhere and watch. You will see lots of things you have been missing. Anyway, I was up in a tree at the edge of the marsh with my notebook jotting down my observations, but I was a bit too far away to hear most of the kids' comments. Frankly, I was dying for some gut-level feedback. Finally, unable to contain myself any longer, I jumped down out of the tree and grabbed the first kid who crawled up out of the marsh at the end of the hour. (Much to his surprise, I should add.) Anyway, he was standing there with his arms spread out and big globs of stuff dribbling down off of him, while I was asking, in succession, "Gee, Billy, how do you feel? What's going on out there? What do you think about all this stuff?"

Billy looked up at me without saying a word, then slowly looked down toward his feet before replying, "It's alright I guess, but what's my mom going to say about my pants?" Sorry, moms, but that was an attitudinal barrier that a lot of our kids came with — a barrier that said it was alright to walk around and look at nature and talk about it, but not to get too mucked up in the process. Nature was something "out there," and lots of kids didn't expect to have much contact with it.

You know, if anything, twenty years later, it's worse. At least, in those days we figured if the kids were wearing jeans that is what they were for. Not so today. "What? No way, man. I'm not gonna sit down in my Calvin Kleins!" Designer jeans may well be the ruination of us all in this field.

In all fairness though, I should tell you the end of my story about Billy. His response about his pants came during the first day of the program, and by the end of the week we had a different problem. You see, the kids wore the same set of clothes whenever they participated in ACC. They called them their animal clothes, and by the end of the experience they were determined to wear them home on the plane. It had become a macho, biker sort of thing, "Nobody's going to get these off me, man."

The second barrier we had to deal with was a real mechanical barrier. I guess the best way to explain this one is to tell you another story about the kids in the marsh. Whenever I would climb up into the trees to watch, I would see the same arrangement, the same pattern of learners out there. Jim would be standing, ankle-deep, in the center of a group of three or four kids enthusiastically sharing a handful of gluck or something with them. In fact, everyone in Jim's inner ring would be touching one another because they were so intensely involved in what was happening. But then there would be a gap of about a foot — picture another imaginary ring around this inner group of kids, only about a foot farther away — and on this imaginary ring you would always find another two or three kids.

In other words, they weren't quite in there; they had drifted a bit. Surprisingly, these kids were usually *aimed* toward the center of the action, but more often than not their heads were turned away. It was evident that they weren't completely involved in the activity underway.

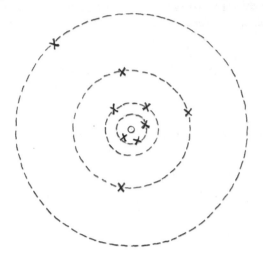

CONCENTRIC rINGS bREAKDOWN

Now picture another imaginary ring only this time about three feet farther out from the last one. On this ring you would usually find two or three kids as well, but they had drifted away entirely. In fact, they were often positioned with their *backs* toward the center of the action. They were either bent over checking out something on their own or gazing off into the distance.

Finally, twelve feet away from this ring, picture another large imaginary ring in the marsh. Invariably, somewhere around this ring you would find some poor lost soul, usually with his hands in his pockets, shoulders hunched, standing on one leg like a heron . . . sinking. He had found some dry little hummock that he thought would hold him, and he was slowly going under. Obviously, he wasn't in the activity (he was just barely in the marsh).

I noticed this same arrangement of kids, and drew it up in my notebook so many times, that I ended up calling it the concentric rings breakdown. Again and again, you would get a few kids right in there with the leader, but more often than not, most of the kids were not fully involved in the activity. They had drifted away from the center of the action. I remember saying to the staff at the time, "Hey, I want every kid in there, in the inner ring, that's where it's happening. That's where the excitement, the participation, the learning is taking place. I am not going to buy having just a few kids in there. I want *every kid* in the inner ring. How can we turn them on to nature if we don't have them in there with us in the heart of the activity?" When you think about it, why is it that even today many leaders still accept a situation like this in outdoor learning where a majority of their learners really aren't participating?

After a lot of agonizing, I became convinced that the concentric rings breakdown was a direct result of the way we approached the marsh in the beginning. It was a mechanical barrier to the kind of learning we wanted to take place. In fact, I became convinced that the concentric rings breakdown was a direct result of the field trip approach, or what I like to call the "follow-me, gather-round" approach to nature education. You know the one, where the leader says, "C'mon kids. Follow me. Gather round over here, I want to show you something. Yackety, yackety, yackety. . . . OK. Follow me over here. Gather round now, I want you to see this. Yackety, yackety, yackety. . . . Alright kids, follow me. . . ."

Every time we took the "follow-me, gather-round" approach we would get a few learners in there with us, but more often than not, most of the learners weren't really in the activity. It was another barrier we would have to overcome in Acclimatization, but before we look at how we did that, let me share one more of those built-in obstacles with you.

The third barrier we discovered was a real physical barrier. You see, we only had an hour in the marsh. And when you only have an hour there, you are going to have to do everything possible to maximize your effort. And do you know what we were doing? We were spending over half our time trying to get them to go in the marsh. I can still hear us cajoling and pleading, "Come on, Johnny. Just a little deeper. It's alright now." For a long time I couldn't figure out why we couldn't just all walk out into the marsh and get going. When it finally hit me about what was going on, I called it the crotchline reaction problem. That's right, and after writing about it in the draft of our first book, I will always remember one of my initial meetings with the publishers to discuss the manuscript. They said, "Steve, can't you call it something else besides that. You're going to offend people." I replied that I would see what we could come up with and they sighed with relief. Anyway, I had another meeting with them about two weeks later, and they asked me if I had come up with an alternative for that "problem" in the marsh. They didn't even want to use the phrase. I replied that yes, after a lot of thought, we had decided to call it the crotchline reaction *syndrome*. And what's funny is that they bought that. "Oh, it's a syndrome, huh?" You know, that is such a great word, you can add it to anything and people will think you know something. So if you are offended, call it a syndrome, but if you have ever waded in to go swimming, you know exactly what I'm talking about. And that's what the kids were doing. They were jumping from dry spot to dry spot trying to overcome the crotchline reaction . . . syndrome.

Well, how did we deal with these barriers? The best way to convey that, of course, would be to take you into a marsh right now, but since we can't do that in reality, let's go into a marsh in fantasy. Picture for a moment a leader and a group of kids standing on the edge of a marsh. The leader is in the middle, with five or six kids holding hands in a line on either side of him. He says something like this: "Everybody hold tight now. We don't want anybody to fall in here.

Acclimatizing continues where the first book leaves off and includes a number of special experiences for building a sense of relationship with the natural world. Well known ACC vehicles and activities such as Earthwalks, Natural Awareness Exercises, Magic Spots, Micro-Trails, Grokking, Seton-Watching, and our Earth Journeys were all introduced here.

Okay, everybody just put your left foot right down in the marsh. Careful now, hold tight!" Why all this chatter? Why can't we just step off and start walking out into the marsh? The reason is probably the most important thing I will say in this chapter: In this work it is just vital to start where your learners are, not where your leaders are. That is probably the most oft-repeated error in western education; we start where the leaders are, not where the learners are. You see, we discovered at camp that our learners thought a marsh was a swamp, and that a swamp was . . . well, you know, bottomless. "It'll just suck you up, man!" And it was undoubtedly full of evil, creepy-crawly things. As leaders we knew that wasn't true, but that was our experience, not theirs. So we had to go very slowly and build on a secure base. This way, invariably, some kid would set his foot down into the marsh, then exclaim, "Hey, there's a *bottom* in here!" And the leader would respond, "Right, our marsh has a bottom in it not very far down. We can walk right out into it."

However, as the group would start walking slowly out into the marsh, another thing we discovered about human anatomy would soon come into play. To be honest, we didn't know how it worked, it must have been a chemical reaction of some sort, but we knew where it was located — about halfway above the kneecap. We called it the crotchline reaction trigger. As soon as the water hit the trigger, you could just see the heads begin bobbing up along the line as the kids reared up on their tiptoes. So we would walk out into the marsh until we hit the triggers, then say, "OK, everybody, let's not go any farther." And the group would usually sigh with some relief, since they thought the weirdo was going to take them all the way in. Instead, the leader would say, "Let's form a circle right here — bring the ends of the line around — so we can hold the most important ceremony in Acclimatization. We call it the Ceremony of the Marsh."

"Everyone squeeze tight now because this is a group effort. That's right, good. You can just feel that group energy. OK. Now let's flex our knees a bit. That's right, just loosen up. You know, I'd like for you to get out of your heads a bit, too. Most people are like worms with swollen sensory appendages on one end — their heads — and that's how they go through life. So let's get out of our heads for a few minutes."

Those words would be the signal for the helper across the circle to start bringing out the black bags and placing them over the kids' heads. Of course, you couldn't begin by asking the first kid, "Kid, do you want a black bag over your head?" No, that wouldn't work. The whole thing had to appear to happen automatically as an integral part of the ceremony. In fact, anything threatening like that would usually get done to the leader first. Naturally, instantaneous silence would encircle the group right behind the placement of a bag over each person's head. And, it's true; you do get out of your head a bit when someone places a black bag over it. You shift down. You feel your pores trying to pull in sensory impressions about what's going on around you. Then the leader would say, "You can just feel the life of the marsh out here calling to you. Everyone take a couple of minutes and just soak up the feeling of all that life around us. I think the marsh really wants us to become totally one with it . . . ," and they would all slowly sit down together in the marsh.

Again, why all this chatter? Because Acclimatization was built on that kind of attention to detail. For example, in the beginning we didn't think about having everybody hold hands tightly. In fact, we must have done that ceremony three or four times before one of the kids decided he wasn't ready to sit down in the marsh with everybody else. So he just let go of his neighbors' hands, pulled off his black bag, and said, "Look at you rummies. You're sittin' in the marsh." Naturally, someone immediately replied, "Get him!", and you can imagine the melee that ensued. After that incident, we decided we had better hold hands *tightly* in the future. And what about flexing the knees? We didn't run across this problem until about a dozen ceremonies later. I will never forget watching Johnny in the circle one day. He had locked his knees, so when it came time to sit down in the marsh, Johnny was pulled down by his neighbors and fell flat on his face. We had to pull him out by his belt. We decided right then, that the next time the kids should *flex* their knees a bit as well.

Over and over again, you will see this kind of attention to detail in Acclimatization. Frankly, there was a lot of hidden structure in our work. That surprises many people. A lot of folks still think ACC meant just going out there and wandering around. Not so.

In our first book we wrote the opening chapters in a "you are there" style. We wanted the readers to feel like they were right in there with the kids. Unfortunately, some people read those beginning chapters — a day (really an hour) in the marsh, a day in the forest, a day in the bog, etc. — but failed to read the chapters in the back of the book that detailed all of the hidden structure, the mechanics and the techniques of learning that those experiences were based upon.

One of the prevalent myths in the field of environmental education in those days was the idea that discovery learning was unstructured learning. After our first book came out in the early seventies, people would arrive at the camp gate, barefoot, with a knapsack on their backs, saying, "We're here." Later, when they found me, they would often add something like, "Hey, what's happening, you're not in the marsh?" I think some folks thought I spent my whole summer submerged in the marsh. Somehow they missed the point that Acclimatization was merely a six-hour barrier-breaking introduction to the natural world, not everything the kids did during their stay at camp.

The point is Acclimatization started out as a special *introductory* program of carefully-crafted, structured learning experiences. And by its very nature discovery learning is structured learning. It is based theoretically on the brain's natural tendency to relate things to one another, to seek meaning, to reach closure. In fact, I believe the structure necessary for good discovery learning is harder to achieve than that needed for terminal behavior learning. I remember staff members in the early days who would say, "Hey, man, can't we just go out there and groove on na-ture?" And I would reply, "Sure, but we're designing a program here to be used in mass education — not for pre-selected, pre-motivated learners — and any time we require kids to participate in a program like that, I think it's incumbent upon us to have some specific learning outcomes in mind." There is certainly nothing wrong with going out there and grooving on nature. I am still all for it, but if our goal is to build some basic ecological understandings, then we need some structured learning experiences to help us get that job done effectively and efficiently in a mass education situation.

Don't misunderstand . . . that doesn't mean there is no room for spontaneity in our work. Years ago one of my colleagues came to me and told me that another staff member had said, "I don't care what Van Matre says, if I round the corner in the morning with a group of kids and discover a Great Blue Heron feeding, we're going to blow off the rest of the activities and watch it." When asked what I thought about this, I replied, "Give the Great Blue Heron ten minutes."

In other words, I felt our program was structured in such a way that a leader could easily take advantage of a special opportunity to instill wonder by watching a heron feed. On the other hand, I felt that the basic ecological understandings we were trying to convey were so important that we didn't want to lose the whole morning. Watching a feeding heron might also be a good *application* of the food chain idea, but since the kids weren't *doing* very much, it would not be that great of a way to instill the idea to begin with. Besides, I had a hunch that it really wasn't the kids who would decide to spend a whole morning watching a heron. . . .

"THE FOUR COMPONENTS OF ACCLIMATIZATION"

1. THE SENSES

I think you can see by all of this that in the beginning Acclimatization came down very strongly on the side of the senses. Candidly, we emphasized sharpening perceptions because many of our campers came to us as if wrapped in gauze. They didn't see much. They didn't hear much. They didn't smell or touch or feel much. In many ways they were insulated, and thus isolated, from the natural world around them. We felt we couldn't help them build a deeper sense of relationship with the earth if we couldn't get them in touch with it.

So for a couple of years, even before there was any thought of a book, we worked on helping the kids sharpen their senses. In a way, I regret it. Why? Because a lot of people today still believe Acclimatization was a program just for sensory awareness. You will even hear them say, "Oh, yeah, we do some of that touchy-feely stuff." But almost from the beginning of any formal programming, we have had other goals in mind. You see, during our early work we found that the kids were definitely sharpening their senses, but we had this growing feeling of uneasiness that something was being left out. Everything around us said that the kids were seeing and hearing and smelling and touching more than they ever had before, and we were pleased, but something was clearly missing. It was like we only had a half a loaf, and we knew the other half wasn't there. We just weren't sure what it was.

2. THE CONCEPTS

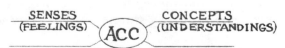

In the end, the missing part was simple. The kids

had sharpened their senses, but not their concepts. Many of our youngsters just didn't have the vaguest idea about how life works here. They were more in touch with it, but they didn't understand it. I remember watching a group one day that was sitting in a circle at the edge of a field. Their leader was talking . . . and talking . . . and talking. Whatever was being said the kids didn't seem to be paying much attention as they were all busy ripping up the grass, methodically breaking up the small sticks lying around, or tearing up any fallen leaves they could find. Finally, I decided I had to see what their leader was talking about at such length and so tip-toed slowly over to the circle. There was a small gap, so I didn't look at anybody, I just sat down and began ripping up the grass right along with the kids. As it turned out, the leader was discoursing at great length about the biogeochemical cycles of the building materials of life. He was going into great detail about the intricacies of the water cycle, the carbon cycle, the nitrogen cycle, etc. At last, when there was a pause in all this, I turned to the kid next to me, who was busy at the time tearing up everything he could get his hands on, nudged him, and whispered, "Frankie, Frankie, what's a cycle?" And he turned, without missing a rip, and replied, "Honda." After all the leader's efforts, Frankie's cycle was still his Honda. I think it was right then that I decided we had to go down both tracks at the same time in our work: the senses *and* the concepts, the feelings *and* the understandings.

Twenty years have passed, and I am still amazed at how many so-called environmental education programs don't focus on basic ecological concepts. They stress the parts of life (water, soil, plants, etc.) and neglect the processes. They emphasize the methods of gaining knowledge (data gathering, sampling, analyzing, etc.), and minimize the understandings. They come up with activities for the easy concepts (habitat, adaptation, protective coloration, etc.) and ignore the hard ones. They insist that outdoor education and environmental education are the same, then do nothing to make sure the kids really understand what supports their lives here and what that means for their future. Even after all these years I am still stunned by the contradictions.

However, let's go back to the marsh. Remember when I said we couldn't call it a marsh because the kids thought a marsh was a swamp? So what did we call it? We called it a bathtub. That's right, a bathtub. You see, our marsh had a very identifiable lip, you could easily stand along one edge of it. We explained it this way: "Kids, this area is like a big, old bathtub. We're standing on one edge or rim and you can see the other rim across the way where the mud has been pushed up by the ice and water over the years. There's higher vegetation growing along that side. You can tell it's an old bathtub because the rim over there is also chipped out in a couple of places and when the lake is high the water sloshes in. . . ." The idea here is that we had to start out with something the kids already had a grasp of in their heads — in this case, a bath-tub — and then build on that understanding. Next, we wouldn't name all the aquatic plants, but say something like, "Inside our old tub there's this huge soggy wet mattress of plants just sort of floating there." And in all the years we used that analogy, we found kids at camp didn't have much trouble identifying with a wet mattress either. If they didn't know personally, someone would usually say, "Yeah, Billy, like yours!" The point is you can see once again how we started out where the kids were, then slowly built on that base.

3. tHE mECHANICS

SENSES (FEELINGS) ACC CONCEPTS (UNDERSTANDINGS) MECHANICS

By this time, I imagine it is pretty clear that in Acclimatization we always stressed the importance of the mechanics of learning. That statement may seem surprising. After all, doesn't everyone involved in this kind of work pay attention to what we know about how people learn? I am afraid the answer is no. To be honest, many of the people involved in nature education don't appear to really know very much about how people learn, or if they do, they certainly don't seem to pay much attention to what they know. Of course, that should come as no surprise either. For the most part, our entire educational system is not based upon what we know about how people learn. In fact, it is most often based upon what we know about how people *don't* learn.

People learn when they take something in, do

something with it, then use it. Sadly, our educational system largely ignores this essential process and is based instead primarily on the taking in phase. The students seldom do much with what they take in (except to try and hold on to it perhaps), and because of the disorganized form in which much of it comes in, they often have little or no use for it later on.

Think back to when you were in school for a moment. Remember trying to talk to someone in the hallway on the day of a test? You would often get something like this, "I can't talk now, man. I'm holdin' on. I've been cramming this stuff in here all night, and I can just barely keep it in there. We'll talk this afternoon after I dump it on the test." Sound familiar?

If we really want to help someone learn something, then we are going to have to pay more attention to how we organize what the learners take in, how we set them up to do something with it, and how we provide opportunities for them to use it again later.

And it doesn't do much good to come up with great learning experiences if the students aren't motivated to participate in them in the first place. The mindless performance of a task, like filling out a worksheet, may yield temporary results, but if you want to produce more lasting outcomes, you have to get the students mentally engaged with their experience. The trick is to achieve a balance between a good task and a good reason for doing it, and thus, in Acclimatization, we always said we wanted to pull our learners, not push them.

Take a close look at many of the outdoor school investigation forms and field study manuals that students have been using for years in this area of education. Sometimes the kids are not even motivated enough to bother filling them in, while at other times they fill them in but don't really understand why they did it. It's just a task someone told them to do. Since there was no intrinsic motivation, there was no real assimilation of the learning. A few hours later and the point of the worksheet is likely forgotten.

Of course, coming up with good tasks and good reasons takes lots of time. If we would give the teachers half of every day free in order to get the good doing ready for the next half day, then maybe, just maybe they could start making the system work. But keep in mind that many teachers have also received very little training in how to conduct highly participatory learning experiences. Chances are these folks are going to fall back on the leadership techniques they learned in the first place, and since you learn best what you do the most with, that is probably not going to represent very much doing for the kids. It's a self-perpetuating cycle. So in addition to giving the teachers the time, some of them are going to need new skills in planning and leading large group activities.

PULLING RATHER THAN PUSHING YOUR LEARNERS

⊕ *Build in appropriate yet appealing rewards*

⊕ *Huddle a group together and in a hushed voice prepare them for what's coming*

⊕ *Promote the next activity with a fast-paced skit during a meal*

⊕ *Arrange a special ceremony that engulfs the participants and sets them up for the task*

⊕ *Challenge the learners to do something different or out of the ordinary*

Since our whole educational system is terribly energy inefficient, all of this, fortunately, will begin changing dramatically as the fossil fuel supplies on our planet continue to dwindle in the years ahead. For example, it is almost criminal what we do now with teenagers in most schools. (In earth education, we believe they are the last, greatest untapped source of energy left in the western world.)

Picture for a moment a human teenage organism sitting in a typical high school classroom. First of all, such organisms consume incredible quantities of food energy. In fact, some of them appear to spend much of their day in eating behavior alone. Then think about how much energy it takes just to move them around (usually from one eating place to another). And, of course, their clothing and recreation represents another amazing energy investment. So what do we do with this energy intense organism today in our educational system? We take this highly charged, energy enriched organism and place it in a room inside an energy intense structure like most school buildings, and what do many of them end up doing in there? *Nothing*. That's right, they just sit there and radiate heat from all that food energy they've consumed. And we don't even have sense enough to use the heat! You would think we would at least suck some of that heat off of them and use it, but no, we often just throw open the windows and let it go. Have you ever seen an infrared photograph showing heat loss in a school building? There are great streamers of red flowing out of every crack and crevice in the structure. We shouldn't call those things schools; we should call them heat factories.

Let's be honest. What lots of kids learn how to do in school is what they really spend much of their time *doing* there, that is, sitting and staring. I call it "S and S" time. In the western world we sure are expending lots of energy in our societies just to teach kids how to sit and stare.

As I pointed out earlier, all of this energy intense system is going to change, and we can take some comfort in that prospect. In the future, as our fossil energy supplies continue to diminish, we will decentralize education again and thus make it more energy efficient; we will use the older kids to help the younger ones, thereby providing better doing for the older and more personal attention for the younger; we will return the extracurricular programs to the kids because the parents will be involved in more labor intensive pursuits themselves (and ask the students to spend more time instead in outside projects applying what they've been learning and thus make better use of a major untapped source of energy here); and finally, we will ask both the teachers and students to do a better job learning fewer things, and thus free both of them up from the warehouses we have created for our young people.

Meanwhile though, what did this all mean for the development of Acclimatization? We decided from the beginning to pay a lot of attention to how people learn in the materials we designed; to tap into one of the greatest sources of energy left in our society — the teenagers — and ask them to help the overburdened and undertrained teachers; to provide for full partici-

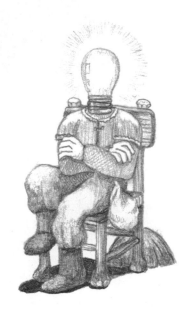

A human being is a 100-150 watt machine. In terms of the heat generated, a person sitting in a room is like turning on a bright light bulb.

pation in our activities by all the students, not just a few; and to provide good learning experiences for people of all ages, and good reasons for doing them.

4. tHE SOLITUDE

SENSES (FEELINGS) SOLITUDE ACC CONCEPTS (UNDERSTANDINGS) MECHANICS

Another of the major components of our work surfaced after we had begun working with the initial ACC pilot. Once again, we had a growing uneasiness that we were leaving something out. I think I first focused on it by name after reading one of Aldous Huxley's essays in which he spoke of the importance of the nonverbal humanities. It suddenly struck me that that was what was missing in Acclimatization.

For the most part our entire education system in western societies is based upon the verbal. It begins as soon as we can get a couple of words out of the little tykes, and it never ends. The further they go in school, the more words they must deal with. Almost by definition, education has become greater and greater doses of verbiage. It is like a great pyramid of words, but it is upside-down. I think if the kids had any idea of what was coming, they never would start talking. "No way. I know what you're gonna do to me."

You know, it is strange, but almost no one in our societies works directly with the nonverbal. And over the years I have become convinced that the nonverbal is just as important as the verbal. In fact, I'm beginning to think it is more important. Take a few minutes right now and think about a couple of the richest moments of your own life. Go ahead, really do it. . . . Chances are good the words faded from those scenes long ago. What remains are the feelings. It is the feelings that endure in life. It is the nonverbal that attracts us — the nonverbal that we long for and return for.

As you have seen, in Acclimatization we decided to focus on the feelings. And we found that there was a very simple thing we could add to our work that would help people sharpen their *nonverbal* skills — skills like watching and waiting, silencing and stilling, opening and receiving. In a word, it was solitude.

THE PERFECT EDUCATIONAL SETTING...

Talking is the most difficult, most complex thing you ever learn in life, and you do it all without school. However, except for walking, when will you ever get such personal encouragement, positive reinforcement, and immediate feedback again? If schools could provide the same stimulation and rewards that most kids receive at home in learning how to walk and talk, our troubles in education would be over.

Chances for participants in our programs to be out there in touch with nature, one on one, in direct contact with the elements of life — light, air, water and soil — unchanneled, unfiltered, unmolded by man. We found that daily periods of solitude would help our learners process what was happening to them while getting closer to the natural world at the same time. They could feel the processes and process their feelings.

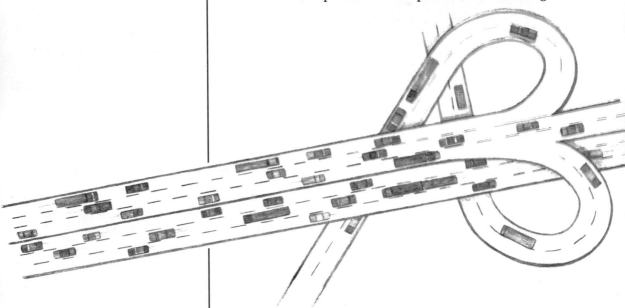

I happen to live outside one of the largest colonies of animals on earth: Chicago. Occasionally, I like to sit back and watch it from my somewhat removed perspective, and I have been struck by the fact that a lot of people in our societies today get up in the morning inside a box, a box that channels and molds and controls the flow of life on earth. Then they go into a smaller box attached to their main box and inside this box they climb into another small horizontally-moving box. Inside these small horizontally-moving boxes they emerge from their attached boxes and get sucked up immediately into one of the gorged arteries that leads to the center of their colony. Like minute packages of fresh energy needed to fuel the workings of the colony's infrastructure they are propelled along inside their tiny horizontally-moving boxes, until they reach the center where they leave their horizontally-moving boxes in underground boxes, then enter tiny vertically-moving boxes which pop them up above the surface again inside tall, gigantic rectangular boxes. Up there somewhere they disembark to spend their entire day inside a little box that is wired for artificial sight and sound. In the evening they reverse the process going from a small box inside a tower to an even smaller vertically-moving box, to an underground box and its horizontally-moving box, then via the arteries to the outlying clusters of boxes to enter a particular attached box and its main box. And for the entire day perhaps the only time they directly touched the elements of life on their planet — light, air, water, and soil — unchanneled, unfiltered, unmolded by man, was when they went out and picked up the evening paper. In the United States today a majority of people spend about ninety percent of their lives inside these boxes. Is it any wonder that they have come to believe that their colonies represent the source of life on earth?

In Acclimatization we wanted to get people out of their boxes and in touch with life again. We wanted them to reach out and touch the earth, to feel themselves as something like microscopic parts of much larger systems. But it is difficult to feel the flow of life if you are too caught up in it yourself mentally. So in a sense we wanted to get them both out of their boxes and out of their heads for a while. When you are too full of your own thoughts, you cannot make room very easily for the impressions of the other life around you. Today, the other creatures of the earth can only be heard by those who work at freeing themselves up to listen.

On the other hand, we also wanted people to spend some time thinking about what was happening to them in our program. Paradoxically, we knew that solitude would also give them a chance to sort things out in their heads. Someone has compared the human brain to a gigantic, gargantuan, gorged warehouse, and working up there behind the counter of your warehouse is a rather bored teenage clerk. In fact, when you belly up to your counter and tell the kid you want to get something out, he may well reply, "You sure it's here, lady?" You answer that yes, you are sure, after all you just put it up there yourself earlier in the morning. Anyway, the kid wanders off in search of your bit of information. He pokes around a while up on the third floor, stops for a dream-stick on the second, and downs a soda on the first. Meanwhile, you are waiting impatiently at the counter trying to hurry him along. Some people scrunch up their faces to do this, others actually knock on their heads, as if they could hurry the process up by pounding on the

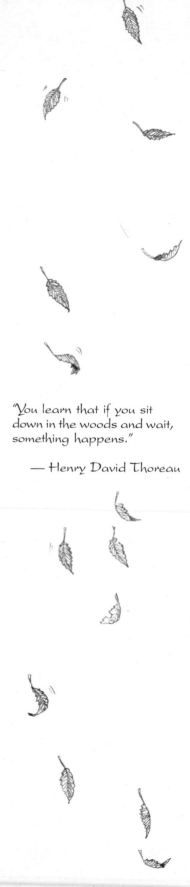

"You learn that if you sit down in the woods and wait, something happens."

— Henry David Thoreau

side of the warehouse. Finally, the kid returns, tosses something across the counter, and at that point people often begin with an "Ahhh . . . ," then end up with an "Oh, no . . . , that's not quite it. It's close; it's probably right next to it in the adjacent bin." How often does that happen to you? You put something in, but can't find it again when you want it. Well, the idea behind solitude time is also to give people a chance to process what is happening to them. I guess another way to look at it is if they can give that kid who works up there in their warehouse a chance to sort things out a bit more in the beginning, he'll be able to find what they want a bit faster in the end.

In short, solitude-enhancing activities became one of the most important components in Acclimatization. Each day we would try to provide some time where our learners could be alone, out there immersed in the flow of life, and thus heighten their awareness of both the other life of the planet and their own innermost thoughts and feelings about it.

"tHE SECRET iNGREDIENT"

SENSES	MAGIC	CONCEPTS
(FEELINGS)	ACC	(UNDERSTANDINGS)
SOLITUDE		MECHANICS

Well, that was Acclimatization: the senses and the concepts and the solitude all wrapped up with the mechanics of how people learn. In the beginning we had the nice diagram you see here to show the various parts. But it just didn't seem to do the job very well. It looked like just another one of those dead dumb diagrams that people put in education texts. Then one day it struck me that it was no wonder it seemed that way, our diagram was missing the single most important part of ACC. It was missing what made the whole more than the sum of its parts, what made it synergistic. We called it the glue of Acclimatization. It was the element that glued those rather disparate pieces together and made them something more than they appeared to be. Sometimes we called it . . . the secret ingredient.

Very simply: it was the magic. That's what made Acclimatization so special. I know, people often think at this point, "Okay, Van Matre, define what you mean by magic." To be honest, I have to tell you I have been trying to define that word for twenty some years, and I can't do it. Everytime I say magic is . . . the words themselves don't seem to have any of it. Maybe that should tell us something about it; words cannot define very well that which is essentially a denial of words. Next, people will say, "Okay, if you can't define it, at least, give us a recipe, like do step one and two and three, and so on, and we'll get it."

Unfortunately, magic doesn't lend itself to the cookbook approach either. I think the best thing I can do here is to share some of the components of the magic with you, and share them in such a way that you can identify with the feelings underlying them, because at the heart of what magic is all about you will always find individual feelings. So this overview is not meant to be an all inclusive definition, nor criteria, nor recipe, but merely an explanation of some of the components of what we call the magic in our work.

The first component I would like to share with you is one that was triggered for me by Haim Ginot. In one of his books, Haim said to give a child in fantasy what you cannot give in reality. As an illustration, he told a story about a mother who had promised her daughter a birthday party. Oh, the little girl was excited. She could hardly wait until next week came so she could have her party. Unfortunately, when her birthday arrived she was sick in bed and could not have her long-awaited celebration. She was just crushed. Regardless of the explanations, she could not understand why she wasn't going to get to have her party. Her mother was heartsick. She knew her little girl didn't understand why she couldn't have her party, but she didn't know what to do. Then suddenly, she had an idea. She went into her daughter's bedroom and asked, "Why couldn't we have a party right here? I mean, just the two of us. Who would've you invited if we could've had the guests come?" And the little girl replied, "Sally, Susie — maybe Billy," and so on. So the mother brought in chairs, placed name tags on them and asked, "Okay, where do you want everyone to sit?" And the little girl decided, "I want Sally right next to me, and Susie over here, and Billy down there. . . ." And in the end, with the help of all the usual trimmings, the mother and her daughter had a birthday party that by all observations was just as successful as if they had had all the kids out in the front room. In

other words, the mother had given her child in fantasy what she could not give her in reality — a birthday party.

Let me share another story with you to take this idea a bit further. A camp director told me years ago that he was sitting on the porch of his dining hall one morning after breakfast looking down towards the lake, drinking his last cup of coffee, as camp directors are wont to do at that time of day. Meanwhile, the kids were back in their cabins doing their morning chores — brushing teeth, making beds, that sort of thing. Suddenly the director looked up to see one of the counselors and a bunch of his boys walking down along the lakeshore. It looked like they were dragging something along on strings. Finally, he went to the door and called down to the counselor, "Say, Bill, could I see you for a minute?" When Bill came jogging up, the director inquired, "What's going on down there? This is clean-up time." But before Bill could reply the kids had tuned in on what was happening and came running up behind him. "Oh, wait a minute, Chief. You don't understand. We have to do this every morning." "Do what, guys?", the director replied. "This is cabin time. You're supposed to be back in your cabin making your beds and stuff. What do you have there anyway? It looks like blocks of wood with strings attached to them." "Oh, no, Chief. We have to do this every morning. It's really important. We're walking our dogs." "W-a-l-k-i-n-g y-o-u-r d-o-g-s?", the director asked very slowly.

Here's what had happened. This particular group of ten and eleven year old boys had gotten together and told their counselor that they wanted him to go into town on his day off and get them a dog. "We want a dog, Bill — a mascot for our cabin." Unfortunately, Bill had forgotten about this until he was driving back to camp on his next day off, when it suddenly hit him. "Oh, my gosh. I bet those kids think I'm going to bring a dog back to camp today. Oh, boy, I should've cleared that up before I left. They've probably been waiting around all afternoon expecting me to show up with a dog." Obviously, Bill was feeling pretty bad and didn't know what to do. He could just visualize the kids' disappointment. Then suddenly he had an idea. Bill drove into camp the back way, went to the shop, cut up some boards and attached strings to them. He thought, "This'll be even better. Every kid can have

In our work, the medium is the magic.

his own dog." And that's what they did. For the rest of the summer, right after breakfast every morning, Bill and his kids walked their dogs down along the lakeshore. "Good boy, Spot. That's a good boy. C'mon, now." Like the mother in the first story, Bill had given his kids in fantasy what he could not give them in reality — in this case, a dog.

I guess the only way in which we would differ with Haim here is that we would drop the word child from his line. We would say, give anyone in fantasy what you cannot give in reality. Why is it that we insist on acting like fantasy is just for kids when we all know that isn't true? Fantasy is one of the richest parts of living. In fact, there's evidence to indicate that people who *cannot* fantasize are frequently institutionalized. Fantasy is an important part of a full life.

One word of caution: in our work we found we had to be very careful with fantasy. We couldn't use so much that it got in the way of the reality. We could add just a pinch of fantasy to get the magic, but not too much. One of the problems we have noticed over the years is that some leaders experience the fantasy that we have woven into an activity, then let themselves get caught up in that, adding even more as they go along. Pretty soon the fantasy dominates the reality. Please don't let your magic overwhelm your message.

For example, the staff at one of our early Sunship Earth pilots wanted to find a way to encourage the participants to keep their cabins cleaned up and prevent "energy leaks" (leaving lights on, water dripping, etc.). So they invented a phantom-like being, the "Cosmic Beam," who would inspect the cabins and leave notes posted on the walls to remind or reinforce where appropriate. Anyway, the participants that season responded with a lot of enthusiasm, and they were obviously inspired to keep their cabins cleaner and the energy leaks sealed. However, the staff "got off" on the excitement themselves and soon the Cosmic Beam — in a colorful costume complete with lightening bolt sash and black hood — began making appearances at meal times to deliver the notes. Next, the staff staged elaborate chases through the dining hall to capture the Cosmic Beam to get at its amazing secrets on saving energy and conserving resources. Suspense, slapstick comedy, complex plots, and great props made for extremely entertaining skits. Once the Cosmic Beam even crashed through a paper barrier riding an intricately decorated bicycle while being chased by a steady stream of staff dressed in full military regalia. The kids loved it . . . but the meaning was lost. The magic had overwhelmed the message. How did they get back on target? The students, unknowingly, helped. The Sunship Study Station began receiving letters that instead of thanking the staff for all they had learned, related how funny it was to watch them chasing the Cosmic Beam.

The second component of the magic is probably going to sound like it doesn't belong with the others. But ponder over this story for a while. I think it belongs in a special way. One day the same camp director in the previous story was walking across his athletic area when he heard some kids yelling and carrying on over among the trees at the edge of the field. When he got closer he could see the kids jumping up and down and shouting to someone up in one of the big trees. Finally, he saw that a platform had been built up there, but guess who was sitting on it with soft cushions behind him and soft drink in hand, leaning back on the tree trunk? Right, the counselor. So when he got over under the tree, the director called out, "Say, Bill, could I see you for a moment?" After Bill climbed down, the director said, "You know, building a tree house is a neat project for the kids, but how come you're up there and all the kids are down here?" Once again, though, the kids tuned in as well and ran over to explain, "Wait a minute, Chief. You don't understand. . . . We built it for him."

So what has that got to do with the magic? I want to take a few moments here and deal with something that hardly anyone in our societies mentions. Everyone talks about giving. "You gotta learn how to give, kids." But hardly anyone in our societies talks about taking. I want to emphasize the importance of taking. I mean, think about it, who would all the givers give to if nobody took? I'm convinced that being a good taker is an important component of the magic. Like Bill said, "Hey, man, I've been crawling up there and sitting in that treehouse every afternoon for a week."

Do you know what I think most of us would've done — me, too, I'm afraid? The kids would've come to us saying, "Hey, man. We wanna build you a tree house." And we would've replied with something like, "Great idea. C'mon, I'll show you how." In other words we would have gotten in between the givers and the gift. We would have cast our shadow over their vision. Bill was a good taker though. When the kids came to him with that proposition, he said, "Great idea. I'll sit down over here under the tree. Tell me when it's ready." Don't you see, that way it remained the kids' idea. It was their dream, their vision, and Bill didn't get in the way.

Regrettably, in our field you will often see leaders who confuse the giving with the taking. We believe there is a time and a place for both, and both can add magic to life, but we must be very careful not to confuse the two. As Bill realized, sometimes *not* giving is the best gift.

At this point, someone is probably asking, "Couldn't you say receive instead of take?" No, I don't think so. Receive is too passive. We need active takers. You see, I believe as strongly as I believe anything that kids want to give, but in our complex, high energy societies many of them cannot find takers. In fact, chances are good that not very far away from where you sit, as you read these words there is a kid crying, crying because he doesn't have anyone to take what he has to give. I believe kids genuinely want to give, but sometimes they can't find anyone to take what they have to offer. Many of them may have already quit looking. For these kids we need active takers — leaders who will just reach out and say, "Hey, wow! For me? Thanks," not passive receivers. In any event, being a good taker adds a special kind of magic to life . . . and it rebounds

to both parties. I guess the bottom line is: both a good taker and a good giver be.

The last component of the magic that I would like to share with you is based upon something that happened to me at Disney World. Yes, Disney World, and I can just hear someone out there thinking, "Unhuh, he goes to Disney World, hey. Not very woodsy." It's true, I do. And I know, it is a big hunk of plastic, and I have some troubles with that myself, but it is rich fantasy.

Anyway, for almost twenty years I spent my winter holidays in the everglades. I usually took some of my graduate students or some of our Acclimatization Associates and we went down to the national park and holed up in the backcountry, worked on some projects and recharged our natural awareness batteries. Of course, on the drive down through central Florida we stopped at Disney World (sometimes we stopped on the way back too). Well, the first year it opened something happened there one evening that I will never forget. It was a highlight of my life. Oh, the staff say, "Steve, don't tell people it was a highlight of your life. They'll think you haven't had much of a life." But it was, believe me, it was. . . .

It happened one evening at the end of the Christmas parade. Folks, if you have not seen the Christmas parade at Disney World, then someday you simply must. It is not just two people and an old horn, it's a parade's parade. In fact, I usually send the people I am with away to watch from somewhere else. You see, as far as I'm concerned they put that parade on for me, so if I am watching it by myself instead of chattering away with someone else, it is just me and my parade.

I usually like to stand right on the corner of the main street where it joins the square. That way the parade comes out of the square to my left and turns down the street on my right, and in the process, I have a wrap-around experience.

You have to understand also that I get terribly caught up in this parade. "Wow, there's Mickey. Ohhh . . . here comes my favorite, Donald Duck. And, oh boy, the bear's gonna shake my hand tonight." You can probably tell, it is about all they can do to keep me out of the parade. In fact, one of my own fantasies in life is that the last anyone will ever see of me is one night I'll just join the parade. People will say, "We don't know what happened. The last we saw of him, he was headed for the castle."

Anyway, there I stood, all wrapped up in my parade one evening when I looked over and saw something happening on the other side of the street that I'll never forget. Over there, the parents were standing up on the edge of the sidewalk while their kids were down off the curb in the street itself where they could get good closeup views. By chance I happened to see one of the fathers over there, standing up on the curb behind a small boy, getting something out of his coat pocket. It looked like a long white card about 4″ high and 12″ wide. I was so lucky to catch this, and I have no idea why I even held on to see what he was doing. You have to understand, I was really torn at this point — Santa Claus was rounding the corner behind me right then.

For the Christmas parade Santa is always on the last float, perched high overhead where everyone can see him as he waves and calls out to those below. Fortunately though, I held on to see the man get the card turned and hold it up surreptitiously behind his son's head, and begin waving to Santa as he pointed at the boy. On the card in large block letters was printed the name, RICHARD. And when Santa rounded the corner, he called out, "Merry Christmas, everyone," and then added, "Oh, hello there Richard!" Well, Richard's mouth dropped open, his eyes almost burst from his head, he doubled up and just shook all over in ecstasy. I am pretty sure he wet his pants.

That is what I call a magical moment. One of those superb moments of pure delight that remain with you forever. I am sure all his days Richard will remember the night Santa called out to him at Disney World. And the whole setup was so simple. His father merely wrote Richard on a card, put it in his pocket and got ready. I never knew that man, but I am sure I know something about him. He cared enough about magic that he wanted to add it to his son's life. And believe me, he did.

You see, I think there are opportunities like that out there most every day for all of us. The trick is not to pass them by. Like that father we need to keep one part of our head on the lookout for ways of adding magic to people's lives.

Several years ago one of my graduate students went to Disney World over the holidays. John had been so captivated by the Richard story that when it came time for the parade, he positioned himself where he thought I had been standing when it happened to me. Just before the parade began he looked over and saw a man and a small boy nearby. Saying to himself, "I've just got to do this," he edged up to the man and whispered, "What's the boy's name? We'll write it on a card and hold it up so Santa will see it and call out to him." Well, this father looked at him like a piece of fecal material had suddenly appeared on the sidewalk. He was actually hostile. He just couldn't deal with that idea.

After the holidays my grads were gathering for our first class, and while we were waiting on everyone to arrive, I asked people to share their recent adventures. As you might suspect, John just sat there looking kind of grim, and when it came his turn he said, "Van Matre, I just want to tell you, this stuff you do doesn't work."

Following John's explanation of his experience at the parade, I replied, "But John, that's exactly why we have to work. That's why we have to do what we do. For that father the magic had withered. We have to help such people renew and refresh their sense of magic again. Magic has to be nourished and cherished to keep it alive. If you are not careful, it will dry up on you. And in our work, we believe magic should be the constant companion in the adventure of living and learning."

You know, when you boil away all the verbiage in education and get down to the essentials, living is learning, and learning is living. The two are inseparable. For us, this is the greatest journey of them all — life — and we are on it right now. We're convinced it should be a joyous adventure, and we think a little magic is a vital part of making it so.

Okay. *That* was really Acclimatization; the senses, the concepts, and the solitude wrapped up with the mechanics of learning and infused with our secret ingredient. That's where we're coming from. . . .

"Magic should be the constant companion in the adventure of living and learning."

CHAPTER **THREE**

EARTH
EDUCATION...
THE
WHYS

"ACCLIMATIZATION BECOMES EARTH EDUCATION"

So where are we going? After years of frustration with the environmental education movement, but with good feelings about the soundness of our Acclimatization experiences, we have started over. Earth Education aims to help people build an understanding of, appreciation for, and harmony with the earth and its life. And anyone can set up an earth education program. Anytime. Anywhere. It doesn't have to be one of ours either. People can create their own. We just hope they will try to make it a genuine earth education *program*. But let's back up for a moment before we analyze earth education in detail and look more carefully at what happened to Acclimatization. . . .

I think I decided that it was time to change our name when I was driving down a road in New Zealand one afternoon and passed a sign reading, "Entering the Western Acclimatization District." I almost wrecked the car! I screeched to a halt, threw it in reverse, and roared back up the highway. Sure enough, that's what the sign said. I didn't know it at the time, but a century or so ago there were Acclimatization Societies established in several parts of the world. However, in a sense, their intent was just the opposite of ours. We wanted to acclimatize people to the plants and animals of the earth, and they wanted to acclimatize the plants and animals of the earth to the people. So after a decade of similar incidents with the old title, in 1984 Acclimatization became earth education.

Notice that I capitalized Acclimatization in the sentence above, but not earth education. Why? Because we also wanted to broaden our base. We wanted people to see earth education as an alternative to the ineffective, moribund environmental education movement, not just a product of our making.

For us, Acclimatization had always been a program, not an approach, but for many folks in the seventies Acclimatization became a generic term for sensory awareness activities. Frankly, we were unhappy about that development almost from the beginning because we thought we were trying to go down both paths at the same time — the senses *and* the concepts. But the most powerful image in our work turned out to be the "immersion" experience, an idea that people latched onto to the exclusion of everything else. For years, we fought to counteract that image and maintain the quality of what we felt was a carefully-crafted introductory program for building a sense of relationship with the earth. In the process we even tried to regain control of the term (a course of action almost never taken, for most people are usually overjoyed when their product goes "generic"), because we felt our intentions and efforts were being diluted, almost trivialized in the field.

Suddenly, there were people we had never even heard of before out there offering Acclimatization workshops and college courses, while others were setting up programs and obtaining grants in the name of ACC. When we objected, the leaders in these situations were often openly hostile. One director in the northeast replied that they were highly skilled at offering workshops themselves, and if we didn't like it, then they just wouldn't sell our books in their network of centres anymore.

Later, people from government funded projects would come to our workshops, then go back and write up our experiences for their own subsidized collections of supplemental activities (often never mentioning their source). This is no exaggeration. We have whole file drawers full of such examples. Sometimes their versions were good (occasionally even improved), but most of the time they would streamline and simplify the activities to the point that they lost all of their excitement and magic. Pulled out of context, the objects of our pride and joy, as it were, now appeared naked and diminished.

I am not talking about your everyday blindfold walk here, or some other such widely used activity of multiple origins, I am talking about activities that were very clearly ours (that were often still called by the titles we gave them). Take "Rainbow Chips," for example, from one of our Earthwalk sets.

The original setup for this activity, the initial bit that hooks and pulls the learners in, goes something like this. . . . As an integral part of a larger Earthwalk, the leader explains to the participants at the end of one activity that a couple of days ago she rounded the corner of the path just ahead and there was this incredible rainbow arching across the sky. Then suddenly, without any warning it just shattered, right before her eyes.

When she rushed over to where chips of the rainbow were tumbling from the sky, she discovered that the colored pieces were falling onto objects of their own color, where they were quickly absorbed. Luckily, she managed to collect a few of these chips before they disappeared, and she wants to see if the participants can find objects now that exactly match their colors.

Of course, this explanation, along with an array of colorful sample "chips" from the paint store, is offered with a twinkle in the eye, but as you can imagine, it serves its purpose well.

Now, take a look at how this same activity was described later in a typical curriculum guide:

rAINBOW CHIPS

1. *Cut up pieces of colored construction paper into small squares.*

2. *Pass them out to the students.*

3. *Have them find things that match their colors.*

Sounds pretty dull, right? Stripped of its magic and taken out of context, it just sits there, stuck to the page. You can imagine what it would do for the kids.

Of course, lots of college students over the years were actually assigned tasks like this in their various outdoor courses. Instead of designing activities to accomplish specific objectives, which takes considerable time and energy, they were often encouraged to "adapt" activities to their needs, and gradually these "quickies" made their way into various collections.

The students were not the only ones who got confused either. A science education professor, writing for one of the journals in the field at the time, complained about attending a session at a conference where the leader asked a group of the participants to use some boards to figure out a way to get across a marsh (their hypothetical roaring chasm). When the prof asked the leader what the purpose of this was, he responded that it was "acclimatization" (the general theme of the conference). As you might suspect, we had never even been informed that this gathering was

In the past we sometimes felt that our Acclimatization materials were both the most widely used <u>and</u> abused materials in the field, but in all fairness, some people were just trying to figure out how to do something with them in their own setting and situation (with not much help or guidance from us). And to be honest, all of our concerns about those abuses also scared some folks off. We were so uptight about what was being done with our work, that they didn't feel comfortable with using anything we had developed. In earth education, we aim to explain not only what we think is wrong in the field, but also how individuals can go about building their own programs for their own needs.

taking place. Perhaps if we had been invited, we could have clarified for the leaders the difference between acclimatization and adventure education, and explained to this professor what Acclimatization was really all about before he wrote his article. Such incidents were a fairly commonplace problem for us for a long time. People with all kinds of other agendas latched onto the interest in our work, and, in the process, often turned it into something else.

So why have we decided that we want earth education to become more widely used, especially after all the frustration and heartache we experienced with Acclimatization? First, because we believe earth education offers an important alternative to environmental education, and second, as I have explained earlier, simply because we know that we cannot create enough programs ourselves to meet the needs people have in all their various settings and situations. Of course, we will continue developing model programs that people can either use themselves or examine to see how they were constructed, but to be successful on a larger scale we will have to help people build their own earth education programs as well.

THE EARTH EDUCATION PYRAMID

Okay. Exactly what is earth education? Earth education is the process of helping people live more harmoniously and joyously with the natural world. To help explain its components we have developed a structural logo, a three-sided pyramid representing the WHYS, WHATS, and WAYS of earth education. And the three points on each side of the pyramid represent one of the three components of either the WHYS, the WHATS, or the WAYS.

pRINCIPLES OF eARTH eDUCATION

tHE "WHYS"

PRESERVING

nuRTURING tRAINING

pRESERVING

We believe the earth as we know it is endangered by its human passengers.

nURTURING

We believe people who have broader understandings and deeper feelings for the planet as a vessel of life are wiser and healthier and happier.

tRAINING

We believe earth advocates are needed to serve as environmental teachers and models, and to champion the existence of earth's nonhuman passengers.

tHE "WHATS"

UNDERSTANDING

feeLING pROCESSING

uNDERSTANDING

We believe in developing in people a basic comprehension of the major ecological systems and communities of the planet.

fEELING

We believe in instilling in people deep and abiding emotional attachments to the earth and its life.

PROCESSING

We believe in helping people change the way they live on the earth.

tHE "WAYS"

STRUCTURING

We believe in building complete programs with adventuresome, magical learning experiences that focus on specific outcomes.

STRUCTURING

IMMERSING rELATING

IMMERSING

We believe in including lots of rich, firsthand contact with the natural world.

rELATING

We believe in providing individuals with time to be alone in natural settings where they can reflect upon all life.

Let's begin with the reasons for our existence. Each of the three "WHYS" of earth education — preserving, nurturing, training — has two facets: *Preserving* the earth for ourselves and for itself alone, *Nurturing* people who have been deprived of a healthy relationship with the earth and enriching those who have not, *Training* individuals who can make changes in their own lives and preparing them to serve as teachers in helping others.

EARTH EDUCATION...
THE WHYS

PRESERVING

"We believe the earth as we know it is endangered by its human passengers."

Some years ago I was interviewed by a television reporter before a speech I was giving in the south-western part of the U.S. They set up the interview at the edge of the desert outside my motel, and the first question the reporter asked was, "Professor Van Matre, how can the people here prepare to live the life they've become accustomed to in the future?"

I laughed and said, "They can't. Look around, you live in a desert. There's no way they can live like this in the future." I went on to explain that their city existed as a sort of energy fungus feeding off of the fossil fuel and fossil water that people had pumped out into the desert.

You see, the water they were using was being sucked up out of ancient underground aquifers, deposited there over thousands of years, and their consumption now lowered the level of that source at the rate of several feet each year. In addition, much of the energy they were using was fossil sunlight that was captured millions of years ago by plants that had subsequently been molded underground by time and pressure into gas and coal and oil. So it was this combination of "old" sunlight and "old" water that actually supported them.

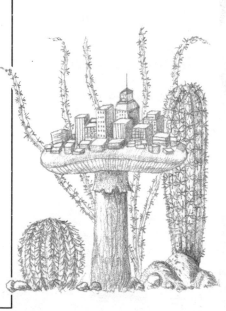

Anyway, the television reporter seemed rather startled by my comments, and said a bit defensively, "Well, we're concerned about environmental problems, professor. Our station has been running a 'Beat the Peak' campaign."

When I asked what in the world that was, she explained that they had been urging people to spread their water use out over the day instead of following the traditional pattern, thus "beating the peak" times of demand. I am afraid I chuckled openly at her example.

"When you look out the
other way toward the stars
you realize it's an awful long
way to the next watering
hole."

— Loren Acton
The Home Planet

She just did not seem to grasp that beating the peak had nothing to do with what I was saying. How can people be so completely uninformed, even misled about what supports their lives? How can anyone possibly believe that "Beating the Peak" offers an answer for our environmental crisis? (By the way, the camera supposedly malfunctioned during the interview and my comments didn't make the evening news. Surprised?)

The point of this story is that so few people really understand how life functions here that it is almost impossible to carry on a conversation with them about many of these issues. They just don't have the vaguest idea of what you are talking about. Consequently, they ricochet off of your remarks into completely extraneous areas, and the conversation becomes very difficult, almost painful to pursue. I remember one man telling me, after listening to the fossil energy and fossil water story, that what he was worried about were all those empty holes we were creating underground. He thought the whole surface of the continent might just collapse.

Earth education exists to preserve the extraordinary richness and biotic health of the third planet from the sun, by changing the perspective and habits of its most dangerous passengers. Unfortunately, many of the human residents of the planet seem unaware of how lucky they are to live on board such a wondrous vessel of life circling a nondescript little star in the corner of an immense spiral galaxy. Powered by the energy of the sun and awash with the liquid of life, their vessel glimmers like a lost oasis in the voids of space. Millions of other forms of life share the oasis with them — from jelly fish and giraffes to kangaroos and gnus; from microscopic creatures in the soils and seas, to towering plants and leviathan animals; from beetles without number to species without names.

Tragically, the health of this marvelous oasis in the universe is now in great danger. The human species of life has become so pervasive and grown so powerful and arrogant that it presently threatens much of the other life with which it shares its garden-like vessel. Estimates vary, but many environmental scientists believe that the human passengers on board are destroying the other kinds of life on the vessel at the rate of at least one species each day.

In the process, the burgeoning numbers of humans have outstripped their own food supply in many areas as well, and over twenty million of them die each year of starvation and nutrition-related disease.

Environmentally induced stress and sickness has also reached plague-like proportions. (In the United States alone almost a thousand people a day die of cancer, and there are now more mental health workers than policemen.) And more and more of the planet's energy supplies and building materials are controlled by fewer and fewer of its inhabitants. (The U.S. Chamber of Commerce estimates that by the turn of the century, a majority of the earth's resources will be controlled by 200 mega-corporations.) Meanwhile, many of those in the northern and western sectors of the planet are caught up in an economic system based upon the idea of unlimited growth even though they live in a limited world. (Uncontrolled growth else-where in life is called a cancer, in America, it's called opportunity.)

At the same time, the systems of life that developed on board the vessel over millions of years without human interference are more and more infected with insidious, man-made poisons with far-reaching environmental consequences, and the highly consumptive lifestyles of a small percentage of the current inhabitants has led to increasing disorder and decay at every turn.

I suspect I don't need to belabor this point any further. You probably feel you are already too familiar with the predictions. In many eyes, soil depletion, biological contamination, habitat destruction, and atmospheric pollution have become the four horsemen of our coming apocalypse, and looming over all, the very real possibility of accidental nuclear annihilation haunts our dreams. As Pogo expressed it in a cartoon in the seventies, "We have met the enemy, and he is us."

"Around the globe we can hear the cries of 65,000 children who die daily of starvation, while the adults spend 1.8 million dollars a minute on weapons."

— Matthew Fox

"The major problems in the world are the result of the difference between the way nature works and the way man thinks."

— Gregory Bateson

MAKING PEACE WITH THE PLANET...

Barry Commoner's sequel to his classic, The Closing Circle, examines how we will have to fundamentally redesign the way we produce goods if we are to solve our environmental problems. Once again, Commoner has set the stage for how we must think in the years ahead.

tHE STATE OF tHE eARTH

How bad is it? Here are some of the current observations and predictions about life on the third planet from the sun.

"We believe the '90s may be our last chance to reverse the trends that are undermining the human prospect. If we fail, environmental deterioration and economic decline may begin to feed on each other, making an effective political response to these threats impossible."

— Lester Brown
World Watch Institute

·"Let there be no illusions. Taking effective action to halt the massive injury to the earth's environment will require a mobilization of political will, international cooperation and sacrifice unknown except in wartime. Yet humanity is in a war right now and it is not too Draconian to call it a war for survival."

— Time Magazine
January, 1989

"Each month we now pour millions of tons of poisonous waste into the global water system. Many of our lakes, rivers, and coastal waters have received their mortal wound. The water is undrinkable. The fish and shellfish, if they exist at all, are contaminated. I do not say this lightly. During the past forty years my team and I have spent thousands of hours diving with Aqua-lungs and other underwater devices. During that time I have observed and studied, and with my own eyes I have seen our waters sicken. Certain reefs that teemed with fish only ten years ago are now almost lifeless. The ocean bottom has been raped by trawlers. Priceless wetlands have been destroyed by landfills. And everywhere are sticky globs of oil, plastic refuse, and unseen clouds of poisonous effluents."

— Jacques-Yves Cousteau
The Cousteau Society

"Runaway destruction of our planet's rain forests — 25 to 50 acres every single minute — is alarming. Especially when you consider that half of our prescription medicines come from natural sources like plants — and nearly half of all plant species on earth are found in these rain forests!"

— World Wildlife Fund

> "The birth of a baby in the United States imposes more than a hundred times the stress on the world's resources and environment as a birth in, say, Bangladesh. Babies from Bangladesh do not grow up to own automobiles and air conditioners or to eat grain-fed beef. Their life-styles do not require huge quantities of minerals and energy, nor do their activities seriously undermine the life-support capability of the entire planet."
>
> — National Geographic Magazine
> December 1988

nURTURING

"We believe people who have broader under-standings and deeper feelings for the planet as a vessel of life are wiser and healthier and happier."

In 1974, as a part of a bicentennial project I was working on, I arranged to visit with a National Park Service specialist in historical interpretation. After a brief meeting one afternoon at his office outside Atlanta, I returned to the city thinking my session had not been very productive. However, the phone rang that evening, and I will never forget his final words of advice: "Steve, listen. Are you paying attention? This is really crucial. Now, get this: *What is important is not the way people were, but the way they thought they were."*

In other words, how people act has a lot to do with how they perceive themselves. If we just look at what people did at any given moment, without examining their thoughts and feelings, we will always come away with a one-dimensional image of them. By the same token, if we just present the ecological understandings we have come to know, without dealing with the human thoughts and feelings involved, we will always end up with a flawed perception of what is happening with our learners.

In the field of environmental education, the term ecological understanding has been used for years, but it is also ecological *feeling* that we seek. If ecology is the study of an organism's relations with its surroundings, then for us, a significant part of that relationship must include an affective dimension.

WHYS
nURTURING

"We should not pretend to understand the world only by the intellect; we apprehend it just as much by feeling. Therefore the judgment of the intellect is, at best, only the half of truth, and must, if it be honest, also come to an understanding of its inadequacy."

— Carl Jung

Earth education
provides a new synthesis
between the understandings
and feelings: ecological
feeling.

Ecologists attempt to understand and explain the behavior of energy at different levels of organization in life, but a vital factor in human energy involves our feelings. Consequently, an ecology text or course that does not take into account the feelings of its learners rapidly becomes a shallow pretense. It denies the very wholeness it claims to represent.

a SENSE OF rELATIONSHIP

For most leaders working in the general nature field, there are typically four common emphases, one or two of which dominates their approach.

⊕ *Nature Appreciation* emphasizes making personal sensory contact and engendering an emotional response. (Those with an artistic bent frequently end up here.)

⊕ *Nature Education* focuses on natural history with ample doses of the victorian identifying, collecting, exhibiting methods blended in with some new environmental messages. (These are the leaders most likely to call themselves naturalists.)

⊕ *Nature Conservation* utilizes management techniques and practices in modifying or working with natural communities and systems. (These folks can often be found working for a government agency.)

⊕ *Nature Recreation* combines outdoor pursuits and skills with games and good times in natural settings. (Many of these people have drifted over from the outdoor or physical education movements, school camping or youth work.)

(Various forms of *Social Education* often take root in our outdoor centres as well, flourishing under an array of titles and tradenames, but generally these folks don't consider themselves a part of the nature field, even though incongruously they may refer to themselves as being a part of environmental education.)

Earth education on the other hand exists primarily to help people develop a better sense of relationship with the natural world. Improving upon this personal contact and connection lies at the heart of everything we do. We believe many people have become estranged from the places and processes that actually support their lives. Thus helping them restore a harmonious and joyous relationship with the earth is our most important task.

A Zen master once said that Zen was the art of making the concrete and the abstract one and the same. In earth education, that is what we are trying to do with the understandings and the feelings. We believe wisdom lies in fully grasping our ecological relationship with the earth — using both our head and our heart — and once achieved, such wisdom brings a happiness uncommon in the modern world.

We are convinced that it is unnatural and unhealthy that many people in our societies today have become so completely removed from the actual source of the energy and materials that support them. Most species of life cannot be taken from their natural communities so easily. Extricated from the matrix of their existence, they are like captured fish flopping forlornly on the ground, vainly seeking the familiar. (Visit a zoo and watch the neurotic behaviors of those poor awkward creatures, then compare what you see with what you will find wandering around in any large urban area.) Perhaps we cannot statistically prove that people who are more connected to the earth are wiser and healthier and happier, but common sense tells us that it must be so.

"NOURISHING, ENRICHING, OR HEALING"

I hope you will not jump to the conclusion that this WHY of earth education is merely a justification for adding some sensory awareness activities to things. It goes much deeper than that.

First, nurturance takes time and patience and skill. Earth education exists to reintroduce people to the real source of their lives, to take them away from their congested colonies, to remove them from the artificial canyons of their cities, to get them out of their synthetic boxes and immerse them in the green and growing things that sustain them. But many people may respond to such an introduction like kidnapped children rediscovering their real parents after a long absence. They have been so deprived for so long from a healthy relationship with the earth that it will take lots of patience to begin breaking down some of the barriers with which they have surrounded themselves. For some people the process will be like a gradual, wrenching

TAKING THE AIR...

Although most people may no longer know the real source of their lives, many of them still intuitively seek it out. After a workshop in England, I hiked up onto the moors on a Sunday afternoon with one of our representatives. Looking back down into the valley below I noticed large numbers of people walking along the sidewalk surrounding a large reservoir. At first, I couldn't figure out what they were doing. The reservoir was enclosed by a low stone wall which they could look over, but the water itself was just a flat, unruffled sheet. There were no birds, no clouds, no reflections. I was puzzled by what appeared from our vantage point to be aimless wandering, until my colleague explained that the people were just getting out of the city. They were drawn to the openness of the natural setting — the play of sunlight on water, the smell of growing things on the breeze, the feeling of freedom.

"I went to the woods because I wished to live deliberately, to front only the essential facts of life, and see if I could not learn what it had to teach, and not, when I came to die, discover that I had not lived."

— Henry David Thoreau

metamorphosis, for others it will occur as a natural unfolding. In either case you will have to be in this work for the long haul.

Second, to enrich means to improve upon the quality of something by adding desirable ingredients. If you say that you want to improve upon the quality of people's lives by enriching their experiences, that implies you may only be adding abundance, that basically their lives are okay. But we believe that for many people their lives are not okay, that without lots of firsthand experience with the natural world people grow up just as deprived as children without good nutrition. They are defective in ways they may never know. For these people a sense of relationship with the earth is more than enrichment; it becomes a basic necessity for the long term health and welfare of our entire species. What we must do in earth education then, is provide nourishment for those who have a poorly developed sense of relationship with the earth and enrichment for those who have already developed a sense of relationship on their own.

In common use, nurturing means to give support during the stages of growth, particularly in the early years. But in earth education nurturing means to nourish or enrich where appropriate, and sometimes to heal. You see, we believe that when you really boil it down, education and medicine are both nurturing fields. One attempts to build, the other to mend. It is good that both are beginning to realize the importance of their similarities, for they should have lots to share. And earth educators will need to learn more healing skills in the years ahead because some people have been so cut off in their lifestyle that a healthy relationship with the earth will be very difficult to build without instituting something akin to a healing process first.

Finally, nurturing is also important for us as leaders. The other two "WHYS" of earth education call upon us to sacrifice, on behalf of other life, but nurturing holds out the promise that there is something in it for us as well. Besides, if we want to convey a love for the earth and instill feelings for it in others, than we must continually renew our own sense of relationship with it at the same time.

The Earth Speaks is a collection of the images and impressions recorded by those who enjoyed a special sense of relationship with the earth. Many of the classic passages of nature literature are included in selections from Thoreau, Muir, Leopold, Burroughs, Carson, Olson, Snyder, Abbey, Eiseley, and others. These nurturing passages can be enjoyed by leaders in moments of personal solitude, or shared with others to help them develop a deeper love for the earth and its life.

tRAINING

"We believe earth advocates are needed to serve as environmental teachers and models, and to champion the existence of earth's nonhuman passengers."

When the IBM company decided to get into the personal computer field, they freely gave away the specifications for their hardware to any software firm that wanted them. As a result, when IBM introduced their first personal computer, there were lots of software packages already available for it. In a sense, that is what we are doing with earth education. We are providing both the framework and the tools (plus this instruction manual) and urging people to begin creating their own earth education programs. Although we will continue to create "model" programs ourselves, we realize that we could never create enough of them (nor publish them fast enough) to match up with all the varied settings and situations in which they are desperately needed.

In addition, through The Institute for Earth Education, we are trying to provide both introductory training sessions and an organizational home for all those who want to become earth educators. We believe that such a support system and its ongoing provision for continued training and sharing will greatly enhance the chances that earth education programs will flourish in the future.

In short, we are convinced that environmental education is simply not the answer for our environmental crisis. If we are really serious about the mission of addressing those problems, we must bypass environmental education and get on with the urgent task of building a worldwide network of earth educators. We hope you will want to join that effort.

In case you are not aware of what has been happening, the environmental movement itself is splitting into two schools of thought. In general, "shallow environmentalist" thinking proposes that we can solve our problems by doing a better job of managing the earth's resources and using technology to clean up the resulting pollution. "Deep ecologist" thinking maintains that our environmental problems are symptoms of more fundamental flaws that will require not only major changes in our lifestyles, but in the very nature

WHYS
tRAINING

ECONOMIC ENVIRON-MENTALISM...?

Recent polls indicate that 75% of Americans now think of themselves as environmentalists. We should be careful, however, in drawing conclusions from such figures because for a lot of folks that may mean they simply want their cake and want to jump on it too. Sadly, many of our leaders are now encouraging them to do just that. The following message from the director of the Environmental Defense Fund illustrates the problem rather well: "What the American public wants — some might say paradoxically — is both to expand our economic well-being and to preserve our natural resources and public health. It is up to us as environmentalists to prove this is no paradox and find the innovative ways to do both."

of how we view our relationship with the earth. In
<u>The Turning Point</u>, Fritjof Capra explained the latter
dimension this way:

> *Deep ecology is supported by modern science,
> and in particular by the new systems approach, but
> it is rooted in a perception of reality that goes beyond
> the scientific framework to an intuitive awareness of
> the oneness of all life, the interdependence of its
> multiple manifestations and its cycles of change and
> transformation. When the concept of the human
> spirit is understood in this sense, as the mode of
> consciousness in which the individual feels connected
> to the cosmos as a whole, it becomes clear that
> ecological awareness is truly spiritual.*

PRINCIPLES OF DEEP ECOLOGY

1. The well-being and flourishing of human and nonhuman life
   on Earth have value in themselves. . . . These values are
   independent of the usefulness of the nonhuman world for
   human purposes.

2. Richness and diversity of life forms contribute to the reali-
   zation of these values and are also values in themselves.

3. Humans have no right to reduce this richness and diversity
   except to satisfy *vital* needs.

4. The flourishing of human life and cultures is compatible with
   a substantial decrease of the human population. The flourishing
   of nonhuman life requires such a decrease.

5. Present human interference with the nonhuman world is
   excessive, and the situation is rapidly worsening.

6. Policies must therefore be changed. These policies affect basic
   economic, technological, and ideological structures. The
   resulting state of affairs will be deeply different from the
   present.

7. The ideological change is mainly that of appreciating *life
   quality* . . . rather than adhering to an increasingly higher
   standard of living. There will be a profound awareness of the
   difference between big and great.

8. Those who subscribe to the foregoing points have an obligation directly or indirectly to try and implement the necessary changes.

> — Deep Ecology
> George Sessions and Arne Naess

Although I have already referred to the cornucopians as some identifiable group (just as I singled out the "supplementalists" earlier), in reality, each of us probably embodies both of these patterns of thinking — the managing environmentalist and the restructuring ecologist — operating within us. Perhaps they are the yin and yang of our field, and we will have to learn how to seek harmony between them. At any rate, I hope you will not let the forcefulness of my assault upon the field characterize someone unfairly. There are probably very few pure cornucopians out there (just as there are very few pure supplementalists). So it is not the people we have to watch out for, but the kind of thinking that influences all of our decisions from time to time.

In The Institute for Earth Education we see ourselves as an educational voice in the broad deep ecology movement. (Although we take no organizational stand on the merits of any particular approach within it.) And we sincerely believe that much of what is called environmental education today is now dominated by shallow environmentalist thinking. Consequently, one of the primary reasons for the existence of earth education is the need to rise above the miasma of environmental education that currently obscures perception in our field. The reality is, environmental education in its present form no longer holds the promise we had all hoped for in its beginnings.

Once again, I know this is strong stuff, but I think we know what we are talking about. In the institute we monitor the journals, newsletters, bulletins, and proceedings from dozens of organizations in our field around the world, and it is quite clear from just a casual examination, that many people still view environmental education as some sort of all-encompassing term, interchangeable with any number of other appellations for outdoor work or environmental concerns. For example, in a recent newsletter the environmental education

SOCIAL ECOLOGY VS. DEEP ECOLOGY...

Murray Bookchin, a leading proponent of social ecology, is highly critical of deep ecology and its advocates. Maintaining that human beings must necessarily change "first nature" into a "second" or cultural nature, and that our environmental problems are rooted primarily in the current anti-ecological structure and social systems of our societies, he believes we must anchor our concern about ecological dislocations in the social dislocations of our times. Bookchin provides much food for thought in the ongoing debate about the future of the environmental movement.

program for one state was described as including "Map and Compass Orienteering, Journal Writing, Visits to National Historic Sites, canoeing and hiking, as well as a large concentration of science-related topics." And this was a description for the program of an entire state.

It is also clear that our field is now thoroughly over-run by representatives of various management perspectives (be they agency or industry or university based). Since there is probably no chance of rooting out this invasion, we believe the only hope for a significant educational response to our environmental crisis is to provide the training necessary to create a viable alternative movement, regardless of its size.

"PREPARING a NEW GENERATION OF lEADERS"

During the years I spent as an educational consultant, I decided I did not like the word teach very well. For a lot of people it seemed to imply acting upon students rather than with them. Sometimes I still think we need fewer teachers out there and more helpers and builders instead. In earth education we believe the real task of teaching is to create stimulating, focused learning situations for specific outcomes, then figure out ways for the learners to repeatedly apply what they have learned. That is why genuine teaching is such a tough, demanding profession. So when we use the term environmental teacher, we mean a teacher in the sense of working with others to help them learn — not showing, but sharing.

If you are already an earth advocate, we want to help you become an earth educator. But genuine teaching for the earth cannot be performed in isolation. As leaders, we have to play out the role in our own lives personally and be willing to take the stage publicly on behalf of our beliefs. As we see it, there are three facets to the role of the earth educator: teacher, model, champion. They are the hidden side of the pyramid in our logo; the three points the whole edifice of earth education rests upon.

"I am always ready to learn, although I do not always like being taught."

— Winston Churchill

We believe:

An environmental teacher is . . .
    a person who helps others understand and
    appreciate and change.

An environmental model is . . .
    a person who demonstrates forming good
    environmental habits and breaking bad ones.

An environmental champion is . . .
    a person willing to publicly defend other life
    that cannot defend itself.

A basic premise of earth education is that if we can get individuals to *re-cognize* the earth (conceptualize it differently), then they will want to *re-present* it to others (communicate it freshly). Consequently, we have to train the leaders first so they can train others in turn.

Of course, in a sense all earth education programs are training programs for their leaders too. Unfortunately, for many of us our personal environmental bad habits are so ingrained, our understandings and appreciations so weak, our contemporary high energy lifestyles so strong that it will take repeated program "immersions" over time to effect the changes necessary. As a result, it may be more effective and efficient to train some leaders briefly beforehand and let the programs serve as reinforcement for their new understandings later.

When you learn how to do many things in life, you maintain and improve upon your skill through practice, through repeatedly doing something with what you have learned. Due to the high energy state of much of the western world, it will also require extensive program follow-through to keep all of us focused on and practicing the skills necessary for lessening our own impact upon the earth. We believe the intrinsic motivation required can be maintained for many leaders not only by repeating a program experience, but also by attending our earth education conferences, or becoming involved with a branch of the institute and its activities. We will look at The Institute for Earth Education in more detail in the final chapter, so now that we have taken a quick look at the WHYS of earth education, we should move on to spend a bit more time on the WHATS.

"We shall require a substantially new manner of thinking if mankind is to survive."

— Albert Einstein

There are three content areas in our work: the *Understandings*, the *Feelings*, the *Processings*. In other words, an earth education program must include building some basic ecological understandings, developing some good feelings about the earth and its life, and processing those understandings and feelings in specific behavioral changes back at home and school. Or for an easier way of holding onto those parts, an earth education program must have components that emphasize "the head, the heart, and the hands."

CHAPTER four

eARTH
eDUCATION...
tHE
WHATS

eARTH eDUCATION...
tHE WHATS

UNDERSTANDING

"We believe in developing in people a basic comprehension of the major ecological systems and communities of the planet."

To understand life on the third planet from the sun, you have to understand the flow of light energy bathing the planet each day, how it is captured and utilized, and how it powers the great cycles of the building materials of all living things — the air, the water, and the soil. Together, that energy and those materials combine in varying amounts in different places and times across the surfaces of the earth. From the beginning, those variations in the quality and quantity of energy and materials available have given rise to various communities of life whose inhabitants are continually interrelating with one another as they go about obtaining their own energy and material needs. And within those percolating pools of life there is the constant ebb and flow of change. Things change one another, they change their surroundings, and their surroundings, in turn, change them.

In short, all living things draw upon sunlight energy for their existence, and each represents a temporary ordered arrangement of matter interacting with its neighbors. Each builds up, then breaks down as the materials of its own body inexorably crumble over time.

Four key understandings then explain the basic functions of life on this planet: The flow of energy, the cycling of matter, the interrelating of life, and the changing of forms. It is these broad brush strokes that we must focus upon in explaining the big picture of life on planet earth.

"tHE fLOW OF eNERGY"

Sunlight streams through space to bathe the surface of this planet each day. Green plants are the only living

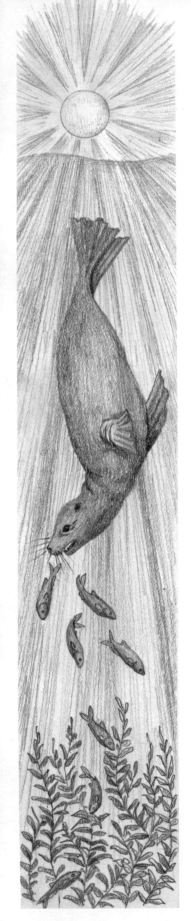

things which can directly capture this energy from our parent star. Through the process of *photosynthesis* they package the sun's energy into molecules of sugar which build leaves, roots, seeds and other plant tissues.

Picture a small green disc in the leaf of one of those plants. It is so small that you cannot see it with the naked eye. Yet that microscopic unit takes a molecule of water and uses a ray of sunlight energy to break that water down into its component parts: hydrogen and oxygen. Next, it uses that sunlight energy in the same way to split a part of air down into carbon and oxygen. Then that tiny disc takes the carbon from the air and the hydrogen from the water, plus a little of the oxygen, and uses the sunlight energy to fuse them together into a carbo-hydrate, and that's sugar, our food. And what is left over? That's the real miracle. Those parts of oxygen given off by the plant as a by-product of the process turn out to be just the right amount you will need to breathe in to release that stored sunlight energy in your own chemical processes. It is not a little more than you will need; it's not a little less. It is exactly the right amount. Just think of it: in the "synthesizing" action in that tiny disc, part of the largest chemical process on earth, you not only gain your source of energy, but you get the oxygen you will need to make that energy available in your own body.

As animals eat and digest the plants, they open these packages of sunlight energy and use that energy to build their own tissues. And the energy travels even farther when another animal (such as a hawk) eats the plant-eater (such as a rabbit). We call this pathway along which energy flows from the sun to plants to plant-eaters to animal-eaters, a *food chain*.

As energy flows along a food chain, much of it is used up and lost. Before they are eaten, plants use up much of their energy just to grow, and animals use up a great deal of the sunlight energy from their food as they move around. Because of this *energy loss*, there is less energy available farther up the food chain, so in most food chains there are many plants, fewer plant-eaters, and even fewer animal-eaters.

Those animals with positions at the ends of such food chains are called top predators. Humans are one of these. However, most humans on the earth today are vegetarians because they cannot squander the captured sunlight energy available to them in the plants by passing it first through an animal. When someone says it is better for you to eat lower on the food chain, she is referring to this loss of available sunlight energy.

In terms of the overall flow of sunlight energy on the earth, life is arranged in pyramids — a broad base of plants supports fewer and fewer animals as the pyramid of captured sunlight energy flows upwards through the plant and animal life in an area. It takes about ten times more plants at the base of a pyramid to support those animals at the next level. And those animals will only be able to support 1/10th of their number in the level above them, and so on. This means that in our energy-intense societies in the western and northern parts of the earth it takes about 100 pounds of grain to produce 10 pounds of beef to produce a 1 pound gain in your weight.

In the beginning, humans, like all other animals, obtained all of their sunlight energy by eating the plants, or eating other animals that had eaten the plants. Later, in learning to use fire, and to harness other animals, they discovered indirect ways of tapping into sunlight energy. In fact, the history of most civilizations on the earth could be recorded in the repeated cycles of discovering, exploiting, and depleting an available energy source.

A major development in the human manipulation of sunlight energy came in the form of cultivating hard-encased seeds. In these small, durable packages of sunlight they found a marvelous device. Now, they could easily store sunlight energy for the future or grind it up and carry it along with them as they spread out over the earth. It would be hard to underestimate the importance of this development in terms of the success of the human form of life.

Since the movement of both air and water on the earth is governed by the energy of the sun, they, too, represent indirect sources of sunlight energy that the human species draws upon. Wind and water power would contribute greatly to the human impact upon the other life here, but it wasn't until they tapped into the fossil fuels buried beneath the surface like so many fatty deposits in the crust of the planet that the impact and pace of human activities began to threaten much of the other life on the earth.

PACKAGED SUNSHINE...

A seed is largely made up of energy-charged sugars for the use of the embryo waiting within it to sprout. Even today, the majority of humans on the earth get most of their sunlight energy from four kinds of hard-encased seeds: rice, wheat, corn and barley.

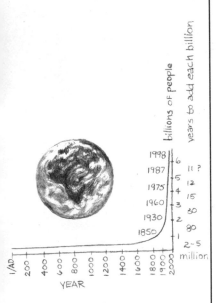

billions of people

years to add each billion

Year	billions	years
1998	6	11?
1987	5	12
1975	4	15
1960	3	30
1930	2	80
1850	1	80
		2-5 million

YEAR

BY WAY OF
COMPARISON...

It is estimated that the average person in Haiti uses the equivalent energy of less than 100 pounds of coal each year, while the average American uses more than 20,000 pounds. Or for another way to look at it the average individual in our society uses the energy equivalent of 200 personal slaves.

The use of fire, the cultivation of hard-encased seeds, the harnessing of the horse for ploughing, and the discovery of fossil fuels would represent major steps in the human "conquest" of the earth and all its life.

Some environmental scientists now estimate that just in the next few decades the human species will destroy one-fifth of all the other species of life that share the earth with them today. Just think of it, one out of every five kinds of life on the earth gone forever.

What has happened is that we have rounded the corner of the j-curve of our impact upon the planet. A j-curve is named after the characteristic shape of a line on a graph charting exponential growth (i.e., 2, 4, 8, 16, 32, etc.). Once you round the corner of the curve in such situations the whole thing just explodes. For example, if you took the sheet of paper these words are printed on and doubled its thickness just forty some times, you would have a stack of paper that reached — the moon! And that is what's happening with our impact upon the life of the earth. We have rounded the corner of a j-curve and the whole thing is beginning to explode.

We are like Faust who sold his soul to the devil for temporary knowledge and power. We are riding this great resurrected wave of old sunlight energy, refusing to believe it will ever end. Now, while there is still time to prepare, when we already know our artificial wave is cresting, we have opted to gamble everything on discovering yet another "source" of sunlight energy to exploit. Instead of getting ready for the future, we have chosen temporary power in order to hold on to our incredibly consumptive lifestyles for just a few more decades.

You see, we are riding up there in the upper lounge of the 747 airplane while most of the other people in the world are consigned to cargo. Is it any wonder that we are resented as we toss back a few scraps of sustenance while at the same time commandeering for ourselves many of the supplies stored in the hold as well? Today, Americans alone consume almost half of the world's resources.

"THE CYCLING OF MATTER"

The basic building materials of life are hydrogen, carbon, oxygen, nitrogen, phosphorus, and sulphur. There is a limited amount of these materials on the earth, so they must be reused over and over again. For millions of years they have been taken from and returned to the air, soil and waters of the earth by all living things.

The soil is the great nourisher of plants, providing the nutrients which they need to grow. The soil would be useless if there was no way of getting these nutrients back after they had been taken up by the plants. But the soil is constantly being replenished, for there are millions of bacteria in each handful of soil that break down waste matter and dead animals and plants, and thus return to it the nutrients which those things contain.

Imagine for a moment that you are holding a big handful of really rich soil. Do you know what you would be holding in that single handful? About 5,000,000,000 living plants. That's right, there are about five billion microorganisms in a single handful of rich soil. Not only that, there are about a million animals in that same handful — eating the plants right in your hand — and about a hundred miles of root hairs holding it all together. Another way for us to look at it would be that there are as many living things in a handful of rich soil as there are people on the earth. The point is, that ain't dirt. That is the placenta of life here. All living things on the surface of the earth go from soil to soil. And so the *soil cycle* continually recirculates the building materials of life.

Water is the liquid of life on earth, and it is moved by the heat of the sun through an enormous cycle of its own. Water evaporates from the rivers, lakes and oceans of the world and rises into the sky where it condenses into clouds. From there the water falls back to the earth in the form of rain, snow, sleet or hail, only to flow once again into the rivers, lakes and oceans of the world. Occasionally, the water takes a detour through a living thing. Evaporation and the filtering action of the soil are important purifiers in this *water cycle*, the largest physical process on earth.

"I bequeathe myself to the dirt, to grow from the grass I love; If you want me again, look for me under your boot-soles."

— Walt Whitman

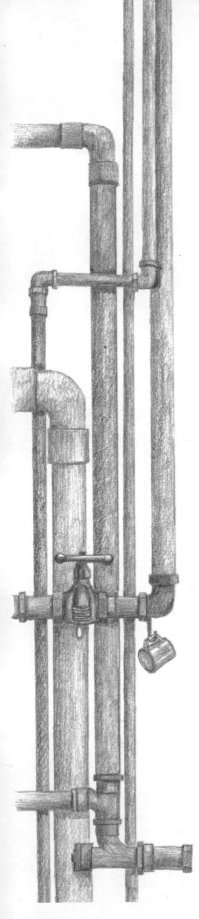

Several years ago a developer outside Chicago invited a group of our faculty to come in and help his company become more environmentally aware on one of their projects. They were in the midst of developing a subdivision for several hundred families, and as you might suspect, they brought us in after they had largely built it. Anyway, they gathered us up one morning inside a model home around a tabletop mock-up of the entire development. All the "boys" were there — the developer, the PR man, the architect, the site engineer, the landscape architect, etc. And one by one each person took the pointer and explained his role in this undertaking. After an hour or so of this commentary, the leader finally asked if we had any questions. I replied, yes, "Where can I go to the restroom?" I sensed immediately that this was not the first question they anticipated answering after their presentation, but I learned some very important things that morning about such questions in our societies. The first thing I learned was that people tend to lower their voice when answering this question. In this case, he whispered, "Down the hall there." The second thing I learned about how we respond is that people sort of lean toward you when answering, and the third thing I learned is that everyone else in the room acts like they don't hear what is going on.

But when I asked what would happen to "it" when I flushed the toilet down there, everyone looked away, especially when I added, "I mean it's mine, I feel some sense of responsibility for it." Even my colleagues hung their heads in embarrassment as if to say, "Why do we let him out?"

However, I was actually quite serious. One of the major problems we face in our societies today is that people take no sense of responsibility for their own waste. "I don't know, man; it goes away," has become the common response to my query. In the case of toilets, I think it was Garrett Hardin who said it best: "Every capitalist becomes a socialist when he flushes his toilet."

A friend of mine tells a similar story about visiting a modern sewage treatment disposal facility. He said there were about twenty people and a guide, and they followed the sewage out of the sewage lines into the holding and settling ponds, then from there into the first and second filtration stages. Finally, when they reached the last part of the process there was a massive wall of pipes and right in the middle of them a huge faucet — suspiciously, with a metal cup dangling from it. Of course, the guide made for this like it was a magnet, asking gleefully enroute, "Does anybody want a drink of water?" Everyone backed up a couple steps, looking rather uncomfortable, except for my friend, who said, "Sure, why not?" However, before he drank the cup of water, he turned to the group and asked, "Does anybody here think that in your entire life you've ever had a drink of water that was *not* something's sewage before?" Chances are good that no one ever had because the waters of the earth have been cycling through the systems of life here for millions upon millions of years.

Parts of the air surrounding the earth are also used over and over again as well. With every breath, animals of the earth exhale carbon dioxide. As we have seen, plants use this carbon dioxide from the air when they make sugar in their leaves, and a by-product of the sugar-making process is oxygen, which the plants release into the air. Thus the *air cycle* provides for the exchange of essential needs of life for both plants and animals.

As perambulating creatures of the land, we live at the bottom of a marvelous layered ocean of air. (Actually, we shouldn't say that we live *on* the earth but *in* it.) The uppermost gases provide a protective shield filtering out the sun's more harmful rays, while far below each of us processes over 2,000 gallons of other gases each day.

Stop right now and focus on your breathing for a moment. Go ahead, take a deep breath and just hold it while you read these words . . . how far down the page do you think you will get? You can go for some time without food or water, but only a few moments without air. However, our gas-gulping habit is so much a part of us that we don't even think about it — taking around 17,000 gulps a day, every day of our lives.

Together, these three great cycles of the building materials of life explain how the matter of the earth is reused — re*cycled* if you will — by all living things through eons of time. It also suggests the potential long-term effects of synthetic, man-made substances

"With every breath you inhale a thousand billion billion atoms. A few million billion of them are long-living argon atoms that are exhaled within a second and dispersed with the winds. Time mixes them and has been mixing them for a long time. Some of them may have visited Buddha or Caesar, or even earlier paid a call on the man from Makapan."

— Rolph Edberg

that have been indiscriminately introduced into these cycles, often without adequate knowledge of their impact.

Not long before he died, the philosopher, Alan Watts, was filming outside his home in California. He would walk along a mountain trail, stopping now and then to say a few words for the camera. When he came to a small field, he noted a single flower blooming there, then said that most of us would probably exclaim, "Wow! Look at that flower blooming in the field." But Alan turned to the camera and said, "No, we shouldn't say that. That's wrong. What we should say is, 'Wow! Look at that field, blooming that flower.'"

Yes, that is how life really works on the earth. All the conditions and forces of life come together in a field to bloom a flower on a sunlit afternoon.

In North America, there is a small nondescript little creature that lives under rocks and stones in ponds and streams for almost its entire life. For several years it lives an underwater existence — then one day it crawls up out of the water and splits open to release a most delicate-winged two-inch creature that takes flight to dance its dance of love in the sunshine. It doesn't even have a mouth nor a stomach in the usual sense, for procreation, not food, is its goal. And it only lives for a single day. After twenty-four hours, it dies, and the building materials that make it up return to the soils and seas of the earth. Sometimes I like to take a walk on a bright, sunny day and every thing I see is dancing its own dance of love in the sunshine, for that is the real story of life on earth.

Stop right now while you are reading this page and pinch yourself. Go ahead, really do it . . . no one's watching. Now, just hold on to that fold of flesh for a moment. Do you know that chances are good that between your fingers right now there's a speck of a . . . *dinosaur?* That's true! You see, whether it is flower or fly or flesh doesn't make any difference. It is all the same building material, and those materials have been used over and over again for a very long time in the systems of life on the earth.

"tHE iNTERRELATING OF lIFE"

Picture the earth as a ball travelling through space, covered with a film of air, water and soil, warmed and energized by the sun. The quantity and quality of air, water, soil, and sunlight in this film varies greatly over the surface of the planet. And this intermixing of energy and materials has produced a great variety of conditions for life on the earth. From the deserts to the oceans, from the arctic to the tropics, from the mountains to the plains, the available energy and building material varies greatly. In turn, this tremendous variety of conditions is the reason for the dazzling *diversity* of living things on the earth. Each living thing is the result of a unique combination of sunlight, air, water and soil, and so is different from every other living thing.

A *community* is a mixed group of plants and animals occupying a specific area of the surface of the earth. Living things are grouped together in these specific areas because it is in such places that they can best meet their energy and material needs. The actual place where something lives in a community, the place where it meets its needs of life, is called its home or *habitat*. A habitat may be as small as a beetle's tiny crevice in the bark of a tree, or as large as an eagle's vast hunting grounds; a good habitat takes up as much space as the plant or animal needs to find its water, food and shelter.

The role or job in the community performed by a plant or animal is called its *niche*. For example, most squirrels basically occupy the niche of a tree-climbing seed-eater, while a common bat's niche could be described as that of a night-flying insect-eater.

Together, the habitat and niche of an individual plant or animal spells out its "address" and "profession" within its community. Over time living things tend to fill all available habitats and niches in a community, and communities appear to function more smoothly when a variety of living things are all doing their "jobs" within them.

Living things interact within their communities in a variety of ways. Although many things we see may be in *competition* with one another for their energy and material needs, under the surface there is also an

"Man is a microcosm, or a little world, because he is an extract from all the stars and planets of the whole firmament, from the earth and the elements; and so he is their quintessence."

— Philipus Paracelsus

Lest we forget: In life, cooperation makes competition possible.

WILL WE NEVER LEARN...

When someone finally decided to rear some gorillas in a social grouping rather than as isolated individuals, some zookeepers were startled to discover that these marvelous creatures exhibited entirely different behavior patterns. Of course, they did. If you were suddenly captured and hauled off to another planet to be exhibited alone in a cage for the remainder of your life, would you not assume different behaviors than you do here? Even if they brought along a member of the opposite sex for you, would your habits remain the same? How about such simple things as eating, grooming, defecating, copulating, etc.? Life can be known in isolation, but it can never be understood.

amazing amount of *cooperation*. As humans, we live in the macro world and thus tend to overlook the underlying micro world that supports us. In fact, a human being itself is actually an interdependent assemblage of over four billion other microorganisms that cooperate to make it up.

Everything on the earth is somehow connected to everything else. It would be impossible for any form of life to exist by itself. There is a parable in Hinduism which expresses this relationship in beautiful imagery. In this story, Indra, the primeval god, cast his net of life out through the vast blackness of the universe. At the junction of every pair of threads in his net there was a crystal bead, and each crystal bead was a living thing — a perfect living entity suspended by its threads in space. And each crystal bead reflected the inner glow, the radiance of every other crystal bead in the net of life. That's how life works on the earth. Each living thing is a spark of sunlight energy here. And each exists in time for only a brief moment, a mere glistening in the film of life on earth. When it fades, the building materials that make it up flow on through the threads of the net to be taken up and used again by other living things. Yet while it is here, that living thing, that crystal bead if you will, reflects the glow of every other living thing around it.

It should be obvious by now: a thing is not a thing out of the context of its community. "Like a fish out of water" is an old saying that harbors much truth. Things are what they are because of the threads of life within which they operate, because of the intricate web of their interrelationships. To remove something from its surroundings is not only to tear a momentary hole in the web of life, it is to disconnect the strands that make that living entity what it is. The light of its life is dimmed.

Tragically, many of the animals you see in a zoo are probably a bit crazy. Wouldn't you be, under such circumstances? And all too many zoos thus represent crazyhouses in the worst and most accurate sense of that old term.

In a large midwestern zoo practically the only interpretation they had in their elephant house for many years was a series of photos with captions explaining that one large African elephant that had gored its keeper was subsequently chained inside its stall for over 20 years. Was that an elephant that people came and looked at during those years? Of course not. It was a crude and tormented caricature of what an elephant is really all about — a magnificent coming together of matter and energy in an interrelated plains community of Africa.

We speak metaphorically about nets and webs of life, yet in reality of course the interrelating of life here is so complex that we can only really begin to understand a mere fraction of it. Lewis Thomas, that marvelous essayist of life's workings, has suggested that we get together with the Soviets and agree not to launch any nuclear missiles until we completely understand just one form of life here. After all, Thomas says, we should have some idea of what we are risking. He even has a candidate: a tiny protozoan that lives in the digestive tract of a termite. However, as you might suspect, once you start examining this supposedly distinct entity, you discover that it is almost a "composite" of different forms of life itself. The minute tentacles that propel it along turn out to be another kind of life that has joined it somewhere along the way in evolutionary time. Next, you find that the very thing you had been giving this creature the credit for doing for the termite (i.e., digesting wood), turns out to be the function of yet another "guest" bacteria that has taken up permanent residence inside this organism. And so it goes. Thomas suggests that even after a decade of gathering such information and feeding it into a supercomputer, the machine would still print out the message: "Request more data . . . do not fire."

"For want of a nail the shoe was lost.

For want of a shoe, the horse was lost.

For want of a horse, the rider was lost.

For want of a rider, the battle was lost.

For want of a battle, the kingdom was lost.

And all for the want of a horseshoe nail."

— Old Nursery Rhyme

"tHE CHANGING Of foRMS"

Each living thing has a unique combination of features and behavior patterns to solve its problems in gathering and using matter and energy. Each has a built-in strategy which provides it with the means to survive — a strategy which enables it to protect itself, get food and water, and reproduce. These *adaptations* may take the form of special features, such as the webbed feet of an otter, the pincers of a crab, or the thorns of a shrub. In addition, the behavior of an individual may also be part of this strategy. Some trees lose their leaves in the winter, birds often fly seasonally to warmer climates, and many creatures are active at night when they can be protected by darkness.

However, life is not static. Over time, forms of life appear with better fitting features and behavior patterns and replace those which are less prepared. This process is the cutting edge for the success of life on earth: the best fit forms survive over time because those individuals live to reproduce and pass along their improved strategies.

This variety also produces an overall endurance and stability. It assures that something can always take advantage of changes in the surrounding conditions. It seems to guarantee that life will succeed in filling nearly every condition on earth, and out of the earth's tremendous pool of living things, something will always be able to survive and life will continue.

Everything on the earth is in the process of changing. In the lives and deaths of plants and animals, in the tides and winds, in the movement and flow of rock within the earth itself, we witness a dynamic, constantly changing planet in action.

Many changes on the earth occur so slowly that we cannot see them happening. The growth and movement of continents, glaciers, mountains and valleys takes thousands upon thousands of years. These and many other major earth changes require enormous amounts of *geologic time* as their most important ingredient.

Certain changes in natural communities happen in a series of distinct stages. One such change is the *succession* of a new kind of plant which is able to live in the shade or soil built up by another kind of plant. A series of such stages can, over a long period of time, cover bare rock with a forest, and even though we could not witness the whole process in one lifetime, we can often see one or two of the stages of such a change taking place.

Imagine once again that the earth is a large, softly-covered stone turning slowly around and around somewhere out in space. Just think how long it has been rotating around and around like that. This huge, turning chunk of matter — now somewhat cooled and encrusted — has been spinning around like a top for millions upon millions of years. You know how long a top will go once you set it to spinning. Just think how much energy it must have taken to set the earth to

"For time changes the nature of the whole world, and all things must pass from one condition to another, and nothing remains like itself."

— Lucretius

spinning slowly around like this — one complete turn every 24 hours now for several *billion* years.

Billion. There's a term whose meaning it has become difficult to comprehend. We toss that word around so loosely these days. Here is a story I use with kids to help them get a handle on that number:

"Imagine that you could go downtown every night right after school and spend a thousand dollars on anything you wanted to buy. Just picture it — taking off every afternoon and spending a thousand dollars. Well, if you had one million dollars total, you could spend a thousand dollars every afternoon after school and at that rate it would last you about three years. But if you had one billion dollars total, you could go downtown every day after school and spend a thousand dollars and at that rate it would last you about three thousand years. That's the difference between a million and a billion."

OUR COSMIC CALENDAR...

The earth has been rotating in space for several-billion years now. Yet it has been said that if the whole history of our planet could be condensed into one calendar year, man would not appear until 10:30pm on December 31st. And "science" — the household god of the twentieth century — would not surface until the last second of our calendar. Isn't it strange that on the basis of such limited experience we have decided that the future is ours?

Change. It is our measure of the ebb and flow of energy. As we have seen, sunlight energy bathes the surface of our planet each day. It is captured by the plants and taken from them, in turn, by an amazing array of animal life — from a tiny sponge to a gigantic whale, from a minute beetle to a ponderous rhinoceros. In a poetic sense, it is the upward push of captured sunlight energy that overcomes the downward pull of gravity, and produces the earth's colorful kaleidoscope of living things.

However, the natural, inexorable tendency of energy is to move from a highly ordered and concentrated form to a disordered and dispersed state. If you have ever seen someone die, you know how quickly the human organism can collapse. Once the flow of the sun's energy is cut off, the "light" literally goes out of a person. All life is an intricately ordered pattern of matter, and it depends on a continual source of sunlight energy to maintain its highly ordered arrangement.

If you have ever had an old car, you know that you cannot let it sit unused for very long either. Only through the rather constant infusion of old sunlight energy (gasoline) is its imminent deterioration prevented. If you leave it alone for just a few weeks, you are likely to return to find that your once cherished and trust-worthy vehicle has suddenly become a pile of junk.

> "There is no such thing as a free lunch."
>
> — Barry Commoner

A FADING DREAM...

This is also why the nuclear power industry has been called a fading dream. When you add up all the energy expenditures necessary (including the recovery and preparation of the fuel, the protection of the sites, the storage and maintenance of the wastes, etc.), it is becoming more and more apparent that our energy return (on our energy invested) is not that great. Any time someone starts spouting energy statistics at you, remind him to subtract all the energy (direct and indirect) that will be required to obtain and utilize the energy he has in mind.

This inevitable, inexorable tendency of all life and all ordered states of matter has been called the *entropy law*. It is nothing new. It has just been given a name.

At some time in our education we have all heard the statement that energy can neither be created nor destroyed, that it is just transformed from one state to another. What this means is that we can't get something for nothing. That is why so many energy projections and discussions today are erroneous. They don't take into account how much energy we have to expend in order to get the energy out that we desire and transport it to where we want it. Remember all the talk about how we were going to squeeze energy out of the rocks in Colorado, and all the projections about "synthetic" fuels we were going to utilize in the future? What happened? Someone forgot this basic premise: it takes energy to get energy.

Okay. That is the first law of energy. But for us, it gets worse. The first law of energy says that you can't get something for nothing, but the second law (the entropy law) states that you can't even really break even.

We have been told that the quantity of energy in equals the quantity of energy out, but that is not the case with the *quality* of the energy. Energy always flows from a concentrated high-quality state to a less concentrated, low quality state. For example, in every link of a food chain some of the rich concentrated energy becomes dispersed heat energy given off by the plant or animal as it goes about its life. Living, growing organisms constantly radiate heat energy into their surroundings, and much of this energy is of such low quality that it cannot be directly utilized again by living things. We must never forget: matter cycles, energy flows — and it is a one-way trip.

Change. It is not always for the better. The entropy law is important because it explains why no matter how hard we try we cannot create more order than disorder in the world. Every time we expend energy in trying to order the world we end up losing more high-quality energy and creating more low-quality, *disordered* energy instead (largely in the form of radiant heat).

True, by using up vast quantities of fossil fuels we can create temporary energy-intense order, but in doing so we are creating, once again, a grand illusion. That's one reason why so many urban renewal projects have failed. In a sense, there just wasn't enough continual energy flowing in (in dozens of forms) to maintain the ordered arrangement that had been developed. The trick is to get closer to the present flow of sunlight energy bathing the earth each day, and escape our dependence on so much of the old sunlight deposited in its crust.

For several centuries we have been led to believe that all we have to do is to work faster and harder and that change or progress will be our reward. Now we realize that such "progress" is an extremely relative and largely illusory description of change. In fact, all growth results in less available energy for the future. The trick then is not to work faster and harder, but slower and smarter.

We simply cannot win on this one. No matter what we do we are going to create more disorder in the world than order. All we can hope for is to get closer to a balance. "Growth" has been the ritual incantation of the past; in the future, the term will be "balance" as more and more families, businesses, communities and countries attempt to stabilize their energy and matter use in sustainable patterns.

Energy Flow. Cycling. Interrelationships. Change. To better understand our place in space, and comprehend what that means for us, we attempt to freeze it in time and scan it line by line like this. But lest we forget, this can never really be done, for the earth is constantly in motion, in flux. Obviously, even these concepts are arbitrary choices, mere approximations of how life actually works here. They are perceptual peepholes through which we can catch a glimpse of reality — a reality that must always remain, tantalizingly, just out of reach.

BALANCING
OUR BUDGET...

In the U.S. we have heard a lot of talk in the past few years about balancing the budget, but we have been working on the wrong one. In the future, the budget we will have to balance is the budget with the earth. And I fear future generations will damn us unmercifully for our arrogance and greed in failing to come to grips with this necessity at an earlier point in time.

fEELING

"We believe in instilling in people deep and abiding emotional attachments to the earth and its life."

It may sound strange, but when it comes to the earth, we are going to have to use it to save it. That is the paradox we now face. I think it is going to be impossible to convince large numbers of people of the necessity of preserving natural communities if they have had no contact with them. Yet our species has grown so prolific on our diet of old sunlight that our numbers alone threaten to fatally wound any natural area that receives us. We are like Lenny and his puppy in Steinbeck's novel. If very many of us start loving it, we may kill it. But in this case, if we don't, it will die anyway, killed by those who never loved it (that may be the ultimate paradox).

The solution to our dilemma appears to be to figure out more and better ways of introducing people to the natural world, then proceed to implement them with great caution. However, before we can figure out the ways, we need to determine exactly what we want folks to take home with them. It's one thing to talk about the importance of cultivating good feelings about nature, but quite another to pin down exactly what those feelings are.

After many years of work in this area, we have concluded in earth education that there are four primary feelings we want people to hold: a joy at being in touch with the elements of life, a kinship with all living things, a reverence for natural communities, and a love for the earth.

"a joy at being in touch with the elements of life"

In earth education we want people to revel in their contact with the natural world, and to seek out opportunities there for the joyous affirmation of their own existence. Joy means happiness and delight, and that is what we want people to experience in being touched by the elements of the earth. It is not something they want to escape from, but something they want to escape to. For us, it is that wonderful feeling you get whenever you are out there in a natural setting, a part of the interplay of light and air and water and soil. It is that heady sense of exhilaration at just *being* — right there, right then — wrapped up in all the elements of life around you.

Frankly, it is a bit difficult to explain the feeling of being fully in touch with life to someone who has not experienced it. But once you have known it, it can never be forgotten. Let me share a part of a letter with you from a fellow who heard me speak at a workshop in Britain. I think he describes his discovery of this joy in a way that everyone can relate with:

> *I'd like to tell you about why nature affected my life so much. About three years ago when I was still working as a graphic designer, I decided to get out of the office for a lunchtime walk. I went to a wood and as I switched off the car engine and got out, the silence overwhelmed me. I went back every day for a year, walked exactly the same path, and day by day I became absorbed, content, delighted, totally at peace. I even found myself laughing out loud at the birds. I let myself go. I lay in the dirt, hugged trees, cried, everything. Most of all, I found that peace I had when I was a young rebellious kid lying on my back on those long summer days of the school break. Having got that feeling back after years of absence, I wasn't about to let it go. Everyone should have it. . . .*

> *Unfortunately, the feeling did go. Gradually, the pressure of work kept me in, the noise of the office, the chaos, the sheer human anxiety and excitement of a big corporation rolling over in its sleep. I didn't forget the feeling though which lasted solidly for about three months. It haunted me. I knew I had to search for it again — get myself into a situation where I could feel it again — so I planned to change my career. It's taken nearly three years to do it. It's difficult trying to explain to family and friends why I should "throw away" a promising career when it was just about to blossom (I'd been offered a partnership in London and had just accepted) because of an abstract feeling I couldn't put into words! It nearly broke me. Financially, it did.*

"The truth is not the truth unless it's felt."

— Archibald Macleish

"Steep thyself in a bowl of summertime."

— Virgil

I tried to get a job as a ranger (without knowing anything about the work really — without "knowing" about nature as defined by the scientists), but was rejected. I stormed into the office and asked to see the head ranger. In the end, they gave me the break, and I just finished the summer as a seasonal ranger. I'm in the field now, but the ground is unfamiliar, and I feel something of the disappointment (though not nearly so much) as I felt when I realized my corporation was in the business of satisfying producers and not concerned about art. I was naive, but shall continue to fight for those principles.

Things are better now, I am working voluntarily with the national trust and nature is on the agenda. Maybe that feeling will come again. I know I must relax. I feel like a homing device searching restlessly for the way ahead. . . .

Two things are particularly significant for us I think in this description. First, the writer was alone when he became aware of the pure, cleansing and releasing sort of joy that can arise from being in touch with natural things. In other words, the kind of joy we are talking about here does not lend itself to a group exercise; to fully experience it, you have to be out there at some point by yourself.

Second, notice how he described his boyhood recollection of "lying on his back" those many summers ago. I think that is important too, because it is that intimate contact with the life of the earth that seems to produce the most profound joy. It's nature face to face, nose to nose, eyeball to eyeball.

At the close of one of my speeches, I urge people to get out there and breathe deeply of the day, to experience their connectedness with the earth — the play of sunlight on water, the smell of a summer day, the song of a field in flower — to know the tingling sensation, the exquisite rapture of melting momentarily into the flow of life. Ah, yes . . . why don't you toss this book aside right now and go out there somewhere yourself and get a hit of it.

"a KINSHIP WITH
aLL LIVING THINGS"

One of my scholars shared a story with us in class about a problem his undergraduate professor posed each year for his students. He asked them to imagine themselves in a car driving along a road bordered on one side by a steep cliff and on the other by a sheer dropoff of some distance. There was just enough room on the road for two cars to pass one another. Nonetheless, the students were supposed to be driving fairly fast when they rounded a corner and spotted ahead of them a rabbit sitting in one lane and a couple of worms crawling across the other. It was obvious they would have to choose to run over either the worms or the rabbit. They just couldn't stop in time. In those few moments available which of these forms of life would they decide to kill?

The idea was that most people would swerve instinctively to kill the worms and spare the rabbit. Supposedly, because the rabbit represents a "higher" form of life, and one more like themselves, the students would feel more empathy with it than with the worms.

Anyway, it's an interesting supposition (and there is probably more to the story), but I would suggest that just because we might instinctively avoid the rabbit doesn't mean that our instincts are justified, nor for the best. Perhaps it is just such instincts that we have to overcome. While it is true that we may feel closer to those "passengers" on our vessel who are most like us, we must work at getting beyond that limitation. And we must extend our concept of rights to include not only the other things that share our life vessel, but to the "communities" of life in which they live, and eventually to the earth itself. In earth education, we want people to feel a bond with all forms of life. For all living things have common characteristics: they process energy, they grow, they reproduce, they face dangers, they do best when the conditions of their homes are most suited to their needs.

Each living thing on the earth is a spark of sunlight caught in a vast web of life. And each living thing is connected through the energy and material strands of that web to every other living thing. In reality, we are one family, a carbon-based family of life with many threads and a multitude of members.

A SENSE
OF RESPECT...

Rachel Carson, author of the environmental classic, <u>Silent Spring</u>, would take the microscopic creatures that she had collected in a vial of seawater and get up during the night, often in the wind and rain, to return them when the tide was once again at the appropriate level. Although she had brought them to her lab to examine them more closely, her feeling of kinship with these minute specks of life would not allow her to return them to the sea at a time when their chances of survival would be lessened.

ALL IN THE FAMILY...

By every definition we are kin to every creature on earth. Even the plants are our brethren. Did you know that only a single atom differentiates the energy-capturing chlorophyll from the energy-carrying hemoglobin of our blood? One is held together by a bit of magnesium and the other by a particle of iron. In other words, the key to energy in both plants and animals differs by only a single, minute speck of matter.

"What we most need to do is to hear within ourselves the sounds of the earth crying."

— Thich Nhat Hanh

It is that feeling of family, that most powerful of human bonds, that we need for the family of life itself. Yes, all the creatures of the earth are our kin. They sleep in our cosmic home, eat at our table, share our air and water, and play with our children. We are composed of their bodies and they of ours. Each of us represents but a brief manifestation of the flow and cycling of life here. In human families we speak of blood lines, but in reality, each of us is intimately related to all the creatures of the earth.

I have a sparrow for a neighbor
and my cousin is an ant,
Today, in my family, we'll welcome a newborn flower,
and tomorrow, bury a fallen leaf.
For I am one with the light and the water and time.

Albert Schweitzer relates that he was journeying up an African river at sunset one evening when the phrase "reverence for life" flashed into his mind. He believed it was his long-sought universal ethic by which all of us could live in harmony with the earth. Although the prose above is mine, I like to think Dr. Schweitzer inspired it.

He explained our dilemma as a conscious, top predator by suggesting that a farmer who cuts down thousands of flowers as he mows his meadow does not, nonetheless, switch off the head of a flower with his walking stick on the way back to the barn. For in the first instance, his killing was out of necessity for his own life, while in the latter case, it was a haphazard, unthinking act of destruction. As Schweitzer put it: "Formerly, people said, 'Who is your neighbor? . . .' Today, we know that all living beings on earth who strive to maintain life and long to be spared pain, all living beings are our neighbors."

Or as a swami in India explained it to a follower who feared they would never meet again: "I am never very far away from you. I am the dust under your feet." Mystics have always grasped better than most the indissolubility of all life. If life is the intersecting flow of energy and the cycling of materials, then it can never be broken. Life dissolves into life.

"a REVERENCE FOR NATURAL COMMUNITIES"

Some years ago at the college they asked me to teach a course called Natural Resources Management. I stalled around for a while, then finally refused. I explained that they had taken two terms, neither of which I could abide, and put them together in the title of one course, and I just would not be able to live with myself if I undertook such an assignment.

What are natural resources, anyway? Stop someone on the street and ask him to name a few and you are likely to get an itemization of "things" — like oil, coal, water, timber, etc. The problem with such listings is that it contributes to our perception of the earth as a cornucopia solely for our use. Instead of seeing the earth as a wondrous vessel of life, we see it as a vast supermarket where we have been turned loose on an endless shopping spree.

Sadly, most elementary school textbooks perpetuate this myth. They contain little lists of natural resources and suggest to the kids that all we need do is to use them more wisely, to conserve *our* resources if you will. They seldom bother to explain the true nature of these "resources," nor suggest that they may not exist solely for our pleasure.

Sure, we can take some of the "resources" from a forest — things like the water that it stores, or the old forest in the form of coal beneath it, or some of its present timber — but if we destroy the community that produces those things in the process, then we have destroyed the real resource — the organizing pattern of life on earth. And I am not talking about one of those crops of trees that are euphemistically called a forest. I am talking about natural communities of life, not those pine tree plantations advertised so appealingly these days in the magazines. Calling one of those crops of trees a forest is like calling a cornfield a prairie.

Management. Now there's a word we need to retire from use when it comes to the natural world. Many people think we can manage the earth. Not so. Basically, we can muck around with it to one degree or another. When we claim we manage a forest, for

> "The 'control of nature' is a phrase conceived in arrogance, born of the Neanderthal age of biology and philosophy, when it was supposed that nature exists for the convenience of man."
>
> — Rachel Carson

A FITTING END...

A natural community is a tightly grouped assemblage of interdependent members. And what is our role? We are an omnivorous top predator with two lobes of our brain that enable us to imagine what is not present, and two small digital appendages that allow us to make our imaginings concrete. Looking over all the other forms of life on earth, we may represent either an anomaly that will destroy itself or life's ultimate joy in knowing itself. It appears to be up to us.

OLD WAYS
PROVE BEST...

For over ten centuries
the rice farmers on the island
of Bali have planted and
irrigated their crops using a
schedule provided by the
priests at the temples for
their water goddess.
Naturally, modern manage-
ment (and banking) systems
encouraged them to abandon
their traditional methods and
build more dams, dig more
canals, plant new high-yield
strains, and use lots of
pesticides and fertilizers in
order to increase their
production. After spending
millions on these new
improvements, only to see
their production gradually
decrease instead, a com-
puter program initiated by a
visiting anthropologist has
demonstrated that the
priests were actually master
ecological planners after all.
It turned out that the old
ways the farmers originally
followed closely matched the
best-harvest strategy the
computer could come up
with.

example, all we are really doing is interjecting our
presence and energy (augmented by old sunlight) into
that community of life. True, we fiddle around with
its systems, we cut or burn or spray, but we don't
really manage it.

To manage something means to control or direct
it. When it comes to natural communities we don't
really do either one. Oh, we can impose our strength
upon them. Like a kid torturing a smaller creature,
we can excise some parts, introduce selectively poisonous
chemicals or perhaps burn off the whole thing, but in
the end, we still can't control it. In fact, every time
the boys think they have figured one of these things
out, they invariably discover they didn't understand
the systems as well as they thought — like the spread of
the painful disease, schistosomiasis in Egypt. When they
dammed the Nile River to provide for more irrigation
and thus bring increased "prosperity" to the farmers,
they also succeeded in creating an ideal habitat for the
freshwater snails that carry this dreaded disease.

Isn't it about time we faced up to it? The word
management should be reserved for dealing with people.
That is what we need to manage. Us. Let's *manage
man*. The systems and communities of life on this
planet apparently prospered for a long time before we
made our debut.

I know. People say we have to "manage" them
now because we have so screwed up the systems and
communities of life here that we have to do things to
rectify the situation. Okay, in some limited cases,
some intervention may be necessary, but let's not
delude ourselves by calling it management and thus
further the arrogance that led us to screw up those
systems and communities in the first place. Let's call
such action cautious intercession or humble support or
some such thing.

In any case, we cannot manage the flow of sunlight
energy here, the largest chemical process on earth.
Not for very long anyhow. Sure, we can use up a lot
of old sunlight energy in the form of fossil fuels to
temporarily allow us to control the flow of sunlight in
some small areas (like Disney's misleading EPCOT
exhibit on food production), but overall, there is no
way we could do it. Nor can we manage the largest
physical process on earth, the cycling of water. Let's

be honest: we are totally subservient to the basic systems of life on earth. They are not ours; we are theirs.

In some ways, the whole idea of management represents a copout. It is a whole lot easier in the short-run to inflict our power upon the earth, than to come to grips with the need to control our own numbers and appetites. Of course, in the long run it is usually someone else who has to pay the price for our arrogance. When it comes to ecological systems, we seldom harvest what we sow.

Fortunately, the concept of management is a rather primitive idea that we will eventually outgrow, but I fear the earth is destined to pay a terrible price for our ignorance in the meantime. And those who bartered away our future biotic health for short-term gains have probably had a building or school or highway or some similar monument named after them and untold generations of their descendants will live off of the fortunes they accumulated in such fashion.

The point is we desperately need to inculcate in our species a reverence for those communities of life upon which we all must ultimately depend. And reverence may be one of the few emotions with sufficient holding power to stay our hand. I don't mean worship. I mean showing a deep respect, paying obeisance to the overall organizing pattern of life here. In case you are into poetic imagery, if the earth is the mother of life, and the sun the father, then the communities of life here should be viewed as their revered offspring.

In short, the people of the earth must learn to see a natural community as more than the sum of its parts. We must see it as a synergistic entity, something to be respected, something not to be trod upon callously and heedlessly, but a place to be treated with great respect and esteem. Such veneration may be our best hope. In a way, we need an earth-centered view of the universe again. Only this time it should be the earth and its communities of life first, and mankind's more egotistical desires second.

NOT ANTI-HUMAN...

Our overconsumption and overpopulation problems as a species have so overwhelmed this planet's natural systems and communities that all life suffers here, not just the other creatures who share our home with us. For a biotically rich and healthy planet, including the human passengers on board, we believe we must focus on the vessel of life as a whole. Concerns over the allocation of seats and the distribution of meals need to be seen in their relation to the life support systems of that vessel, not as ends in themselves. Without a holistic perspective we are doomed to endless quarrels and crises as things get sorted out in one area only to break down again in another. It is for the good of all life that we must pay more attention to the ecological constraints with which we are bound.

"a love for the earth"

After I conducted a session at one of our sharing centres in England a few years ago, the director told me he had been meeting in London at the same time with a couple of representatives from the North American Association for Environmental Education. During their conversation, one of these folks remarked that they had had a lot of trouble in the states with people who thought environmental education was hugging trees and kissing bears. I don't think the director told them that the preeminent tree hugger of them all was conducting a workshop right then at his centre, but it's sad that it became fashionable to refer to those who espouse an emotional appeal for the natural world in such an off-handed way. Indeed, there has been a noticeable tendency among environmental educators in recent years to gain "respectability" in the academic community by becoming more erudite, more empirical, more research-oriented, but in the process, the field appears to have lost much of its mass appeal. (That is probably why a lot of potential environmental educators have found a more comfortable home in outdoor education, for some of those folks at least still demonstrate a bit of passion for the earth and its life.)

Unfortunately, many of our leaders have been unable to provide any clear sense of direction for those whose emotional response would have made them our most heroic of advocates. Of course, we must have studies and serious scholarship, and it is good that educators have some knowledge of their results, but reason alone will not save the earth. In the United States, at least, it seems that the environmental education movement has been influenced from the beginning by academics whose interest was not primarily lodged in the techniques of mass education and communication, but in the disciplines of environmental studies and resource management.

The earth is in trouble not simply because people don't understand. It is more than that. Lots of them just don't care. They have lost a feeling for where they really are. They have literally "lost touch" with the other life of their planet. Like distant friends from their past that they have not seen in many years, they no longer have the necessary strong emotional attachments to sustain them in a healthy relationship with the earth. Their synthetic, energy-filled lifestyles have

"Nature never did betray the heart that loved her."

— William Wordsworth

separated them from the natural systems that produce their sustenance, and isolated them from the natural rhythms that govern their spirit.

If the earth is to survive the onslaught of its human passengers, if we are going to come to grips with the necessity of limiting our reproduction and consumption, then a love for the planet and its richness of life probably represents our best hope. Why love? Because when you love something you will give things up for it, and that is what we must do for the earth. We must sacrifice our appetites on behalf of the future. Why love? Because people are more likely to make rapid and lasting change for emotional reasons than for rational ones. Why love? Because people will fight for what they love much faster and much harder than for what they merely know.

Some people ask how is it possible to really love the earth? I see its beauty, smell its aroma, feel its caress, taste its sweetness, and hear its sigh every time I sleep with it. Is that not love?

Why are such people afraid to say they love the earth in the first place? Have they never wanted — somewhere, sometime — deep down in their chest, just to reach out and give the earth a hug? Have they never had a day so steeped in the earth that they alternated between tingling, surging waves of pure emotion and indolent, melting, drifting moments of pure relaxation? When they felt so alive, so perfectly in tune with the vibrations around them that on one hand they thought they could just burst open like a ripened pod, and on the other they felt like they could just melt away into their surroundings? Is that not love?

Have they never yearned for a place that so captured their feeling they felt less than whole when they were away? Have they never sat in a sylvan glen or trod the desert sand or leaned against a mountainside and thought that such a place was where they really belonged? Have they never walked in the unending surf or watched a sunset over some unknown horizon or paddled in the wilds with the loons and sensed that they were most fully alive when they were most fully in touch with the earth? Is that not love?

"If you love it enough, anything will talk with you."

— George Washington Carver

A DEEPER RELATIONSHIP...

Falling in love with the earth is one of life's great adventures. True, a few people evidently do this through naming and categorizing its parts, through the examination and study of particular pieces, but most people appear to fall in love with the earth through direct exposure to its elements, through firsthand contact with its communities, through rich experiences with its overall panoply of life. Hunters, fishermen, golfers, birders, orienteering enthusiasts and many others share the pursuit of rather singular goals, but most of them speak of their love for the outdoor settings of their pursuits. (It's just too bad that we haven't been able to help more of them also enjoy an emotional relationship with the natural world on its own terms, without the props and tools of their sports.)

We say people love money, or fine clothes, or cars and sports. Why are we so reluctant then to say they love the earth?

In traditional nature study we have all too often dissected and dismembered the real object of our admiration. We have analyzed the parts instead of the wholes. We have focused too much on the skeletal structure and not enough on the flesh. We have spent too much time with the cells and not enough with the communities. Most people fall in love with wholes, not parts; with flesh, not skeletons; with communities, not cells.

Remember when they told us in biology class that we would start out gaining an understanding of the smallest units of life, then gradually put them together to see how they formed the larger units? In reality, many people seemed to have gotten lost along the way, and by the time they finally got a field trip worked into this schedule, many of the students were already so overwhelmed by the detail that we may never again recapture their interest for the bigger picture of life on earth.

By focusing on the pieces of life instead of its processes, we have fostered an egocentric view of the earth instead of an ecocentric one. We have made it easier to love ourselves perhaps, but harder to love the planet that sustains us. When you love another person, your emotional response may be triggered by a small part of the person, but it is the whole that captivates and holds you. In affairs of the heart, it is the whole that you cherish and to which you return.

Other people say that love is based upon understanding, that you cannot love what you don't fully understand. Nonsense. If that were true, the religions of the world would be in a lot of trouble. You often love the unfathomable in life, perhaps in direct proportion to its inscrutability. Besides, you love other people without understanding the functions of their pituitaries or the intricacies of their kneecaps. And although it may be true that you don't really love very many individuals beyond your own species, people often speak of loving specific communities of life they have known. In the end, we believe in earth education that to love the earth is to love ourselves, for we are of the earth and it is one with us.

"A man doesn't learn to understand anything unless he loves it."

— Goethe

Joy. Kinship. Reverence. Love.

Once again we must not confuse our attempt to make sense out of these feelings with the feelings themselves. In that mysterious inner well of our emotions, these feelings intermix like drops of color in water. We can analyze them in their pure form for only a moment or two before they dissolve into the whole to which they belong.

PROCESSING

"We believe in helping people change the way they live on the earth."

When it comes to the real crux of our environmental crisis, I do not think we can blame people for not understanding what's going on. Everywhere they go they are assailed by descriptions of the problems: endangered species signs at the zoo, save the prairie campaigns at the club meetings, raptor rehab programs at the nature centers, environmental degradation documentaries on television, etc., but no one ever tells them that it is their lifestyles that are causing the problems. No one ever says, "Hey, the way you live is the reason for all this. That is the primary cause of our trouble."

The point is that understanding and feeling are not enough if people don't incorporate them into their personal lives. They have to act upon those understandings and feelings in direct ways. So the third "WHAT" of earth education is this processing of the other two — absorbing those insights and relating them to our own lives on a daily basis. And just as in the other sections, we have identified four components that we must focus upon: assimilating understandings for how life works on the earth, enhancing feelings for the earth and its life, crafting more harmonious lifestyles, and participating in environmental planning and action.

"INTERNALIZING UNDERSTANDINGS FOR HOW LIFE WORKS ON THE EARTH"

A biologist friend of mine was giving me a hard time about over-simplifying things a while back, when I turned on him, rather roughly I'm afraid, and said something like, "Hey, you guys have had them for several hundred years, and they still don't understand how life works here. Don't you think it's about time you gave somebody else a chance?"

Tragically, most people today just do not understand the big picture ecological concepts about how life functions here, nor what those processes mean for them in their own lives. And what's worse, there is little doubt, most of our political leaders are ecologically illiterate as well. And whose fault is this? Ours — the educators; we supposedly taught them. The politicians have certainly had an explanation from us about how life works here, in fact, they have had it several times during their education. They probably had it at a couple of points in elementary school. They surely had it in high school. And most of those folks even went to college.

Yet I am confident that lots of our political leaders could not adequately explain energy flow on the earth, or the cycling of the building materials, or the inter-relating of life. Oh sure, some of them might recall a few of the words, but it is doubtful they would know the tune. So what happened?

I think we have to come to grips with it. What we have been doing in school just isn't working for most of our learners. They receive the explanation, but it just doesn't take. They never really assimilate it into their own lives. Besides, how life works here is often presented as something pertinent for other creatures, but not mankind. By the time anyone gets around to tying people into these systems, the course is finished. In earth education it seems clear to us. What we have to do is to go back and start over. We have to help people understand who and where they are and what this means for them in their daily lives. If we ask how many revolutions they have made around the sun, or suggest they just arrived at the school or office or centre

"Nature and books belong to the eyes that see them."

— Ralph Waldo Emerson

"A child educated only at school is an uneducated child."

— George Santayana

on a ray of sunshine that took a sixty million year detour, they should know what that means. Or what does it really mean to turn on a light switch, plug in an appliance, drive a car, eat lunch, buy a paper, or perform any one of hundreds of other daily acts? Can people actually see themselves as a spark of sunlight energy caught in a complex web of life? And can they really grasp that everything they purchase and everything they discard has a profound impact upon that web over time?

As a professor of interpretation as well as environmental education, I often visit interpretive sites to see what kind of work they are doing with their visitors. One of the sites I checked out a few years ago was the Plymouth Colony Settlement on Cape Cod. Re-created to represent the lifestyles and times of the early 1600's, the costumed interpreters go about their chores now much as they would have done back then, while the visitors observe and interact where appropriate. One summer morning I was standing off to one side taking in this interaction when I happened to notice a costumed guide go over and grab up a couple of chickens from the side lot of a thatched-roofed cottage, while another guide came out of the cottage across the lane carrying a kettle of boiling water.

Well, I was prepared for what was going to happen as chicken inevitably headed for the boiling water, but I was totally unprepared for what was going to happen to the people on the street when they figured this out.

First visitor: *"You're not going to kill those poor chickens are you?"*

Guide: *"Well you eat chicken don't you?"*

Second Visitor: *"Yes, but that's different."*

Third Visitor: *"C'mon Johnny, we're not going to watch this. This is mean."*

Everyone walked away. Honest. All that was left in that whole area was the kettle, the chicken, and Van Matre.

However, if that is not bothersome enough, here is the real frightening part of this tale. I got curious about where everyone went, and when I looked

ARCHITECTURAL BIOLOGY...

The Baubiologie movement in Germany aims to help people literally build a complete new life from the ground up, including their homes, furnishings, diets, clothing, etc. Like feng shui, the oriental counterpart of building in harmony with the environment, baubiologist home builders begin by taking into consideration the natural setting, the materials to be used, the spatial arrangement of their rooms, and how these elements will work together in a whole. In short, they build lifestyles, not structures. It is to groups like this that we should look for insights into crafting a more environmentally harmonious future.

(See the note at the end of the Earth Educator's Bookshelf for information on obtaining a catalog of Baubiologie hardware.)

around, I discovered a lot of those folks had gone to *hide* to watch the killing of the chickens. That's right. You could see them peering out from behind the trees down the street. You could see their little eyes gazing out from behind the shutters of the cottages. In a sense, they had become voyeurs on the killing of their own food.

Most people in our societies today have been so isolated for so long from the realities of life here that they simply do not grasp how their own lives are a part of the overall process of life on earth. We have to help them begin peeling away the disguises we have used here to mask the ecological processes that support us.

For many people, their entire lives have been played out as if they were participants in a grand masquerade, and by now they have been absorbed so completely into their roles in this fabricated energy "diversion" that they are almost totally oblivious to the fact that they are acting out their lives in an artificial stage setting. If you doubt what I am saying, go to any shopping mall and spend some time watching the people there. We have succeeded so well in our efforts to conceal the basic functioning of life here that many people may no longer be able to deal with naked reality. When the lights dim, the sound system fades and the props begin to crumble from a lack of the continuous high-energy input required for our grand illusion, will our neighbors be prepared for life beyond the curtain? Yes, but only if we begin helping them unmask now.

To assimilate something means to incorporate it into your own way of thinking. So we need to help our participants see the flow of energy, the cycling of materials, the interrelating of life, and the changing of forms in everything they do. Even though those systems have always been there, we have become so adept at covering them up in our societies that we no longer realize they operate all around us. By removing some of the disguises we have used to mask the actual workings of life, we can help our participants see that humans are not really separate from nature; they have just hidden it away under layer upon layer of artificial glazing.

Fortunately, opportunities for helping people peel away such disguises can be found in most every setting and situation. Years ago at the camp where our work

"By closing the eyes and slumbering, and consenting to be deceived by shows, men establish and confirm their daily life of routine and habit everywhere, which still is built on purely illusory foundations."

— Henry David Thoreau

began, there was a small chalkboard outside the main dining lodge. Each morning one of the cooks would write up what was coming for lunch, then return in the afternoon to put up what we would be having for dinner. By chance, I was watching the cook post the dinner menu one day when I suddenly realized that that wasn't what we were going to eat at all. It was a disguise. So after she had gone back inside, I went up and erased it. On the first line she had written, "Roast Pork." I changed that to "Roast Pig," then decided that still wasn't clear enough and changed it again to, "Roast Hoofed Mammal." The next line became, "Boiled Orange Roots," followed by, "Mashed White Tubers." And the last line read simply: "Tossed Leaves." That is what we were really going to eat that evening — roots and tubers and mammals and leaves. You could do the same thing in your own setting today — at home or school or office or centre — just call your food by that which it is instead of by that which it is not. By peeling away such disguises we can begin to see our connections again with the systems of life around us.

By the way, if you do include "Tossed Leaves," be sure to add in parentheses, "Sprinkled with Insect Repellents." It's true, no matter how many times you wash those tossed leaves they still have residues of insect repellents on them when they reach the table. You see, as you sit reading this right now, there are men in little white coats, in little white rooms creating new substances that never existed before in the film of life on earth. They are creating these new substances at the rate of about a thousand new ones each year, and there are almost no controls on them. The Environmental Protection Agency in the U.S. estimates that about 50,000 of these substances have been created so far, but the E.P.A. has only been able to adequately test 5,000 of them. (And that's using their own definition of what is adequate.) In other words, the folks in those little white coats are creating these new substances faster than anyone is able to test them to see what effects they are going to have on our ecological systems. And massive quantities of these substances have already been released into the film of life here as insecticides, pesticides, and herbicides.

In my opinion, these are the bad guys of the insect repellents. Why? Because we just don't know where they are going out there nor what they are doing. Let me qualify that. We know one place for sure where they are going. Everyone reading this has some of those synthetic, man-made substances concentrated in their bones and fatty tissues, and the amount seems to be growing larger. We just don't know what this means for life. Remember: the systems of life here developed over millions upon millions of years without dealing with these new substances so we are not sure what effects they may be having out there. I guess I should qualify that statement too. Hardly a month goes by that we don't hear about how another one of these substances has turned out to be carcinogenic, and it is now estimated that at least half of all cancers are environmentally induced.

On the other hand, there are some good guy insect repellents as well. What do most of us do when we get our bowl of tossed leaves to the table? Would you believe, we add more insect repellents to them? You see, four out of every five animals on the face of the earth are insects. Four out of every five. And they are out there right now munching away on the plants. It is probably a good thing that we don't hear so well as a species. Can you imagine what it must really sound like out there? Anyway, some plants, by chance, developed things over time to ward off all those insects. And that's what we frequently collect today and put in little jars on little racks in our kitchens to add to our bowls of tossed leaves . . . spices.

Of course, there are other ways to help people internalize their understandings, but disguise-removing has the advantage of hitting home almost immediately. After hearing about my chicken story in a workshop, an elementary school teacher told me that she wouldn't have paid too much attention to my remarks about all this disguise-removing business if it wasn't for something that had just happened to her with her own family. Each day she took her kids home with her at noon and fixed them lunch there. One day recently she had been opening a can of chicken soup when she overheard one of her children say something that seemed to indicate he thought chicken soup came from a can. Fortunately, she decided to check this out. So she turned to the kids and said, "You all don't really believe that chicken soup comes from a can, do you?" "Of course, mom. Chicken soup comes from a red and white can." "No, I mean where it really comes from in the beginning." "C'mon, mom. It comes from a can. Everybody knows that."

At this point she went to the phone, called her principal and told him that neither she nor her kids were coming back to school that afternoon. "We're going to go downtown and get a chicken and make soup." I never did understand how he bought that — "Okay, Martha . . . whatever you say" — but she discovered her first hurdle was to convince the kids when they got to the supermarket that what they were getting was a whole chicken. "Mom, that's not a chicken." "It's naked!"

Then when she got it home and had it simmering on the stove, the kids were saying, "Yecchh! You don't think we're going to eat that stuff, do you? It's dead!" You know, in the end she had to force her kids to take a spoonful of it. And guess what they said? "Gosh, mom. This tastes just like . . . chicken soup." We need more mothers and teachers and principals willing to get the kids involved in peeling away the disguises on their own soup.

Disguise-removing can be a lot of fun; just wander around your house or school or centre asking yourself what various things there really are, and what do they actually represent in terms of energy and materials on the earth (plus their impact on the other life here that shares those with us). But it can also be rather painful at times as well. Try calling milk-fed veal, for example, by that which it really is instead of by that which it clearly is not.

rEMOVING dISGUISES

⊕ Call things by names that reveal their hidden nature or underlying reality (foodstuffs, beverages, appliances, etc.)

⊕ Add explanatory signs and diagrams that illustrate how common items are tied into the ecological systems (toilets, light switches, drains, waste baskets, vending machines, etc.)

⊕ Trace things to their true origins or destinations (hamburgers, human excrement, clothing, etc.)

⊕ Analyze common activities in terms of their actual energy and materials use (recreation, transportation, education, etc.)

⊕ Post "fun facts" that focus attention on the environmental impact of daily actions

"ENHANCING FEELINGS FOR THE EARTH AND ITS LIFE"

If we are serious about our educational mission, our feelings for the earth and its life cannot be left to wither with only an occasional "watering" on a Sunday afternoon, when we put on our tweeds and go for a walk at the local nature center.

As we have seen, in earth education we believe we have to nourish in others a lasting love for the earth, a joy at being in touch with wild and growing things, a feeling of kinship and reverence. We have to get them out there more often, and help them notice the sky and the trees and the wild things whenever they go. We have to help them absorb enough that it comes oozing back out of them in the way they walk and talk and dress, in all the paraphernalia with which they surround themselves. It is not enough to simply intellectualize these feelings; people have to make them an integral part of themselves. It has to be nature in the navel.

In order to accomplish this task we use some tools that may appear familiar at first, like values clarification exercises. In fact, there has been a lot written in the past about the importance of values clarification in the environmental education movement. But to be honest, in earth education we don't want to clarify people's values; we want to help them build some. So we use the term values-building instead of values clarifying. We believe values-building activities can help people process their feelings for the earth and its life, as long as we are very careful to keep such activities grounded in actual experience.

Much of the underlying (and unacknowledged) assumption about values clarification in our field seemed to be that people already held precursors of positive environmental values, and if we could just help them clarify those, everything would be fine. But that may not be true. Many people are so removed from the natural world, so isolated from the real source of their energy and matter, that they may hold very few positive environmental values. Besides, values don't come from discussion. They arise out of experience. In many of the collections of environmental activities that were published in the seventies you got the impression that the authors thought values would spring full-blown into

Make it nature in the navel.

existence if you just spent more time sitting around in the classroom talking about them. We disagree.

A major problem with many of the early values clarification activities that were developed was that they tended to enhance the polarizing effect of group discussion, or succumb to the invisible forces of peer group interaction. As an educational consultant, I saw so many instances of people completing a values clarification activity with what actually amounted to passive, noncommittal, popular responses that I ended up coining the phrase, "a value is not a value unless it's a verb." (That changes, for example, peace to peacemaking or friendship to making friends.)

Another problem in this area appeared most clearly in the work of the "overnight" trainers. I recall one colleague who returned from a conference where he had briefly watched one of the "founding fathers" of values clarification demonstrate some exercises only to become an instant expert himself. He would literally go off to the back room the next day to read over an exercise, then conduct it out front with a group of leaders. Of course, such activities were novel — and everybody seemed to be doing it, like a new parlor game — but when he was finished, what did he end up with? Where had he really taken his participants? What could they do with what he had done (other than become instant trainers themselves)?

Like most useful learning, building lasting feelings for the earth and seeing them transform people's lives (and that's where it counts) will take multiple focused exposure, much reinforcement, and repeated application. We will need a lot more than a couple of one-off exercises prepared behind the scenes moments beforehand then delivered up by an overnight expert.

Properly prepared, and conducted as a part of a larger program, we believe values-building activities can help us strengthen old feelings and kindle new ones, but we must always take them another step. We must ask our participants what they can do to get more of those good feelings into their daily lives.

That is where it's really at in this section, for just as we can peel away the disguises on things people do each day to help them digest their ecological understandings, we can examine their daily routines looking

"To feel beauty is a better thing than to understand how we come to feel it. To have imagination and taste, to love the best, to be carried by the contemplation of nature to a vivid faith in the ideal, all this is more, a great deal more, than any science can hope to be."

— George Santayana

for new ways to enhance their ecological feelings. Even better, we can have them do it themselves. It's easy. Just ask people to make a list of those things they normally do — from the moment they get up until the moment they go to bed — each day during the week (save the weekends for later). Now go through the list with them looking for ideas about how they can do things in a way that will also support or enhance their feelings for the earth.

---

bUILDING VALUES

⊕ Ask participants to recall and relate good feelings they have had in natural settings (favorite wild places, magical afternoons, meeting other creatures, etc.)

⊕ Encourage participants to share natural things and places with others (letters, gifts, mini-trips, etc.)

⊕ Invite participants to explain both their strengths and weaknesses in building a stronger personal relationship with the earth and its life

⊕ Ask participants to publicly champion the existence of natural communities

⊕ Arrange for participants to spend time alone with another living thing in its neighborhood

---

Don't make this too complicated. Look for opportunities to add small natural touches. A touch of something is ordinarily just a light tinge or trace or tint of it. In this case, a natural touch may require a bit more, but it should be just enough to convey the flavor, the feel of natural systems and communities.

There's a marvelous Zen story about an apprentice monk assigned to clean the garden path in a monastery one autumn. When he finished the master came to check on his work, and after looking over the path, simply said, "No, no, that will not do." So the apprentice went back to work, although this time he carefully raked and swept up the fallen leaves from every square foot of space. He removed every small twig, picked up every stray leaf, then sprayed everything with water.

However, once again the master came to look, and left saying, "No, no, that will not do."

On his third try the monk crawled along the path on his hands and knees. He took a small brush and scrubbed each stone. He used his fingers to rake every blade of grass into just the right position. This time when the master came to check on the results, he stood back and studied the whole effect, then walked up the path to where a small tree arched overhead. He grabbed its trunk and shook the tree vigorously to create a path dappled with beautiful orange leaves. "There, that will do," he said.

The point of this story is that the apprentice concluded his attempt with a perfectly cleaned and arranged space, and the master added just the right touch of natural imperfection. The same consideration holds true for natural touches. They are introduced after all the basics have been attended to first. They must never become a substitute for doing your chores.

Natural touches can be added to almost everything you do (and all the objects around you), but to be most effective they must appear to have naturally settled there (wherever you place them) like the leaves on the garden path. They must speak of harmony (of balance and quietude). They must never be jarring, but appropriate to their setting and situation.

Touches are not merely decorative objects, but those things that demonstrate thoughtfulness in terms of introducing natural elements. They are a bit harder to explain than disguises because they seem to cover a wider range of possibilities, but here's a story about interlocking touches on different levels that suggests some of their potential for creating magical experiences.

Upon my arrival at the camp in Oregon last year where we were holding the institute's annual staff gathering, our host, Bill Weiler, explained that my room was located in the infirmary. When I asked if it was any particular one, he simply replied that I would know it. When I entered the infirmary I noticed immediately that the door on the opposite side of the lounge had a rough sketch hanging on it of a face with a long beard like mine and the caption: Wizard's Den. Inside the "den" I discovered a whole series of special touches. Perched on the windowsill was a chunk of lava

containing several small plants; a beautiful golden leaf rested lightly on the table; an Indian pipe decorated the dresser; two marvelous etchings of owls hung over the bed; a special leather-bound copy of <u>The Earth Speaks</u> and a feather waited on the nightstand, etc.

After taking these in, I spotted an egg carton, sitting on a small table under a potted plant, with a label on top saying, "Touches." I smiled, thinking about our activity by that name in the Earthwalks binder, an activity that uses egg cartons containing an array of natural items designed to help people focus on textures. But inside this carton I found some mysterious photographs instead — close-up snapshots of all the touches that had been added to my room.

I went through them, comparing each with the appropriate item and found that I had missed a couple of things — one photo led me eventually to discover a journal on the closet shelf, but strangely its pages were all blank, and there was another photograph that I could neither figure out exactly what it represented nor locate any object like it in the room. It appeared to be some sort of carving — like a small mask inlaid with shells — on top of a short pole.

After breakfast the next morning I returned to my room to discover that the blank journal had appeared on my nightstand with a long piece of cord obviously marking something. I opened it up to that page and found that the following lines had been written there: "I am going to show you my stick, said the old man. For it is a stick that sees. Grandfather, that is nonsense, said the second guard. That stick can see no farther than the emperor's daughter."

On the following day, we loaded up everyone at the meeting and left the camp for an all-day excursion to the institute's northwest office in Trout Lake, Washington. When we returned that evening I was surprised to find that all the touches in my room had disappeared. There was nothing left — just a barren, rather unappealing bedroom in a camp infirmary. However, where the carton of touches had been there was now a children's book marked with the photograph of that strange carving. I was in a hurry and had to rush off after merely glancing through it, but my feeling of being led on a marvelous discovery (another of our principles) was growing. When I got back to the main

lodge, I was also surprised, and a little puzzled, to see that the touches that had added so much to my room now added an extra sparkle here and there in our meeting areas.

The next morning I got up early to read <u>The Seeing Stick</u>. The story of the seeing stick begins with a blind princess in ancient Peking and the emperor's offer of riches for anyone who can help his daughter see. Sadly, none of the forthcoming prayers or potions or pins are successful, until one day an old man appears with a walking stick in which he carves the images of his experiences, and using it he teaches the blind princess how to see through touch instead of sight.

ADDING TOUCHES

⊕ Use natural items to convey a touch of nature (as a part of your furnishings, clothing, vehicles, etc.)

⊕ Add a natural representation to a setting or scene (a picture, recording, carving, etc.)

⊕ Make more natural, more earthy purchases whenever possible (items crafted with care from simple, but durable, more naturally occurring materials)

⊕ Serve natural foods and drinks (fresh, organic, homegrown)

⊕ Bring the outside inside (with unprocessed, undecorated bits and pieces of the natural world)

⊕ Add something natural to your interactions with others (letters, gifts, meetings, etc.)

Now, I was beginning to see more clearly myself. The photo was the top of a walking stick, and sure enough Bill appeared with it after dinner that evening at our traditional evening of awards and remembrances. He presented it to me along with the wish that it be passed along each year to another of our Associate staff members. Bill explained that the pacific northwest native American who carved the stick said that it was a "seeing stick" and would help those who used it to really see the life of the earth. So our Associates will carve their initials and the appropriate year in the stick and pass it along from one to another at each staff gathering.

In the years ahead, the seeing stick (another natural touch for the conference itself) will circle the world, experiencing different languages and lands, but always reminding its recipients to see with their hearts . . . and of the importance of touches.

"CRAFTING MORE HARMONIOUS LIFESTYLES"

At the 1984 World's Fair in New Orleans, the U.S. Bureau of Land Management sponsored a pavilion that could serve as a harbinger of our times. Here are a few of the passages from the script of their multimedia production on "How Water Won the West:"

"There is so much life in the desert. It waits only to be released, to be set free."

"We made all this happen by working in harmony with nature. We brought water to this dry countryside. Now, it bursts with life and productivity."

"We deal with the vast forces of nature — forces that are sometimes unleashed upon us with untamed power. We must walk in harmony with these forces."

"Man and nature have joined together; they have achieved a delicate union that has made them one."

Did you catch some of those disguises, like "working in harmony," or "achieving a delicate union?" To a person concerned about the environment such phrases sound promising, but in this case they are frightfully misleading. For the BLM folks, working in harmony with nature means controlling and taming and manipulating her to be more "productive." From their perspective, nature is viewed as a separate set of forces that we "team up" with for the betterment of our families and the future. We are not a product of nature — grown rambunctious on our diet of fossil fuels — we are a separate entity that has achieved a

"Under favorable conditions, practically everybody can be converted to practically anything."

— Aldous Huxley

"delicate union" with her. In the eyes of the BLM: "Man courted nature. He recognized not only her beauty but the potential of her creations. He took her hand. Each — man and nature — brought their special and unique offerings to the relationship." All this is simply the old man-centered, man-dominated view of the natural world now projected in benevolent and rosy hues, onto an unsuspecting public.

It is sad that none of our environmental organizations were able to counter such propaganda with a pavilion of their own. Of course, they did not have access to public funds to finance the presentation of their messages. Millions of people are now subjected to similar exhibits at our centres and parks and conferences each year. As I noted in the opening chapter about our lost educational mission, much of the environmental movement as a whole is being co-opted by governmental agencies and private industries that have a mission diametrically opposed to a sustainable earth society. If nothing else, "How Water Won the West" could serve as a good example of why we should change the language of the environmental movement every decade or so; for the goals of environmentalism are expressed in terms that are easily adopted by those who do not really share the same vision of the future.

In earth education we have a different relationship with the earth in mind, and the harmony we want to craft does not mean merely imposing our desires upon the natural systems and communities of the earth, but learning to live within their limitations. However, a craft is a special skill or art, while a lifestyle involves a person's whole way of living. Crafting a particular lifestyle then, rather than merely accumulating one, takes both time and determination. Ordinarily, our lives take on the look of one long binge at a never-ending flea market. Trash and treasure are thus intermixed in our lifestyles in a haphazard assortment of routines and rituals, possessions and pursuits. Even when we are motivated, it is difficult to begin sorting out all the accumulated ways we have of doing things — discarding some, saving others, looking for more harmonious alternatives. A lifestyle molds itself to us like an old, well-worn outfit of clothing. We feel comfortable with it, and change doesn't come easily.

What do we mean by living in harmony with the earth? Does it mean that you never kill, that you live

"We may be stewards of our own household, but we are tenants of the earth."

1 + 1 = 5,000,000,000...

I used to tell my undergraduate students, "It's easy for you to be environmentally sound. You don't have anything." The problem is that as we accumulate more in life we find our impact increasing almost unconsciously. And when it comes to procreation, our impact increases exponentially as well. Some years ago someone suggested that perhaps we shouldn't list in the institute's staff newsletter all of the new babies born to our associate representatives, for we may be rewarding those parents for having the greatest impact they will ever have on the earth. In fact, maybe we should be listing instead the names of those who had decided to limit their reproduction.

If each of their offspring marries and produces an average number of children, a typical U.S. couple would be responsible for adding over 7,000,000 people to the planet in a thousand years, while the average world couple would add over twice as many.

in a log cabin, that you don't reproduce, that you own nothing? Of course it doesn't. In practice, what it means is that each of us keeps trying (in our own ways and at our own pace) to lessen our impact upon the earth and its life. For most people, a harmonious relationship with the earth represents something that they will always be working towards, a goal that continually recedes as they gain more understanding and greater feeling for the communities and systems of life here. To live in harmony with the earth is to *be* in harmony with it, and for most of us that will take a lifetime of effort.

So we have to start with people where they are (not where we are as leaders, but where they are as learners). We have to help them begin analyzing their present lifestyles, whatever those may be like, and begin making improvements relative to their personal impact upon the earth. As a leader, the important thing is to get them started, then pull them along with continued reinforcement and support. Remember the old Chinese proverb: "A journey of a thousand miles begins with a single step."

What lies at the end of the road? We are not sure. There is no one way of living in harmony with the earth as far as we know, but that's what makes the challenge so interesting. Everyone has the chance to try new paths — to explore, experiment, discover, create — to build personal models for the future.

You can begin helping others by asking them if they would like to put some real adventure into their lives. Ask them if they would like to know that what they are doing with their lives represents a meaningful contribution to the future of the earth. Ask them if they would like to feel that they are growing personally in rich, new ways. If they answer yes to these questions, then offer to help them start crafting more harmonious lifestyles and sharing their results with others.

In crafting such a lifestyle, a good starting point is to begin analyzing habitual actions, with the intent of breaking bad environmental habits and forming good ones instead. A habit is a patterned response to certain conditions. People grow so accustomed to doing things in certain ways that those ways become unconscious behaviors. In a sense, they do them without knowing that they are doing them. To break these

CHOOSING HABITS...

Naturally, you will want to select environmental tasks that are appropriate for your participants in their setting and situation. Here are three important sources of ideas for what learners of different ages can do to lessen their impact on the planet:

50 Simple Things You Can Do To Save The Earth
The Earthworks Group

The Greenhouse Crisis
(101 Ways to Save the Earth... and How You Fit into the Puzzle)
The Greenhouse Crisis Foundation

Personal Action Guide For The Earth
United Nations Transmission Project

(See the footnote in The Earth Educator's Bookshelf for information on where to obtain these materials.)

patterns they will have to focus upon them first. They will have to stop and see them freshly, then begin looking for alternative approaches.

In The Institute for Earth Education, we have begun spelling out some environmental habits to focus on in each of our earth education programs. Since habits are particular behaviors, they lend themselves to the kind of specificity needed for definite behavioral outcomes. Here are some examples from our upper elementary program, Earthkeepers, which focuses on energy and materials. We hope they will serve as some possible starting points for your own work with others:

"Habit is habit, and not to be flung out of the window by any man, but coaxed downstairs a step at a time."

— Mark Twain

lESSENING iMPACT tASKS

eNERGY (CHOOSE ONE)

☑

☐ Heat . . .
when you feel cold, put on a sweater rather than turn up the heat

☐ Electricity . . .
when you are not using a room, make sure all lights and appliances are turned off

☐ Transportation . . .
instead of getting a ride in a car to somewhere you go often, begin walking or riding a bike

mATERIALS (CHOOSE ONE)

☐ Water . . .
when you are using water to wash your hands or brush your teeth, turn the faucet on gently and leave it running only as long as you need it

☐ Paper . . .
write on the back of writing paper instead of throwing it away

☐ Aluminum . . .
recycle all the aluminum you use

> "When the forms of an old culture are dying, the new culture is created by a few people who are not afraid to be insecure."
>
> — Rudolf Bahro

Fortunately, there are already a lot of other people out there who are trying to live more in harmony with the earth. We should be tapping into their expertise — drawing them into our conferences, writing about them in our newsletters, incorporating them in our programs. Naturally, many of these folks will not be found in the mainstream of education, for they are out there living in simpler ways and thus aren't plugged in to many of our groups and meetings. But the alternative lifestyle movement is alive and well. We have listed some books and periodicals on the next pages that will give you an introduction to this area, but chances are good, wherever you are, there is someone living nearby that you should get to know.

SELECTIONS FROM AN EARTH EDUCATOR'S BOOKSHELF

BOOKS

<u>Seeing Green</u> —
Jonathon Porritt

An excellent overview of the Green Movement, including philosophical and practical suggestions for change. (The most readable book on the list, and the first one to get your hands on.)

<u>The Real Cost</u> —
Richard North

Examines the actual, often hidden costs of everyday items like bread, coffee, hamburgers, jeans, etc.

<u>Person/Planet</u> —
Theodore Roszak

Maintaining that the needs of the person and the needs of the planet are the same, Roszak suggests that our exploration of human potentialities and our search for an enlightened ecology have begun to merge as a force for social change.

<u>The Reenchantment of the World</u> —
Morris Berman

Challenges the mechanistic view of traditional science and suggests a new metaphysics to replace it.

Extinction —
Paul and Anne Ehrlich

A comprehensive, "pull no punches" account of our impact upon the other forms of life that share the earth with us.

Economics as if the Earth Really Mattered —
Susan Meeker-Lowry

A primer of whole earth economics full of numerous suggestions of how the average person can invest in building a new economy.

Voluntary Simplicity —
Duane Elgin

Based on a survey of alternative lifestyles, it contains many practical suggestions for "being personally."

The Turning Point —
Fritjof Capra

A critical examination of the dominant world view along with an introduction to a more holistic, systems-based alternative. (This one could change your life.)

Blueprint for a Green Planet —
John Seymour and Herbert Girardet

Calling itself "a handbook of positive measures and realistic alternatives," it's full of eye-catching, explanatory graphics, and page after page of things you can do to live more harmoniously with the earth's systems of life.

Gaia: A New Look at Life on Earth —
James Lovelock

Introduces the hypothesis that the earth itself functions as a living organism.

Eco-Philosophy —
Henryk Skolimowski

An expose of the fallacies and flaws of modern industrialism.

Quest for Gaia —
Kit Pedler

Lots of practical, nitty-gritty ideas for living more harmoniously on the earth.

Towards an Ecological Society —
Murray Bookchin

A leading exponent of social ecology (as opposed to deep ecology) presents his views for a better tomorrow. Lots to think about.

The Politics of the Solar Age —
Hazel Henderson

A hard-hitting examination of contemporary economics.

Gaia, An Atlas of Planet Management —
Norman Meyers (ed)

Despite the use of the "M" word in the title (and not surprisingly, some of the conclusions that its contributors reach), this is a good sourcebook for graphic data on what's happening to the earth.

Permaculture —
Bill Mollison and David Holmgren

Describes an alternative approach to agriculture that combines landscape design with food production right in your own back (and front) yard.

The Spiritual Dimension of Green Politics —
Charlene Spretnak

Examines the linking of spiritual values with the principles of green politics.

Wilderness and the American Mind —
Roderick Nash

A well-written analysis of the origins and development of American attitudes towards the land and its life. (A must read if you live in the U.S.A.)

Dwellers in the Land: The Bioregional Vision —
Kirkpatrick Sale

An introduction to bioregionalism and its potential for the future.

Nature and Madness —
Paul Shepard

Argues that many of our environmental problems are a result of stunted human development.

Earth Wisdom —
Dolores LaChapelle

A wide-ranging philosophical examination that moves from the sacred mountains of the earth's people in the past, to the present relationship between mind and nature, to suggestions for creating a new sacred ecology for the future.

State of the World —
Lester Brown

An assessment of the environmental state of the world prepared annually by the Worldwatch Institute.

Entropy —
Jeremy Rifkin

A forceful survey of previous worldviews and a convincing argument for a new one.

The Arrogance of Humanism —
David Ehrenfeld

A powerfully written (and controversial) attack upon our blind faith in progress and reason and its destructive consequences.

Deep Ecology —
Bill Devall and George Sessions

An overview of the many past and present facets of this growing movement, plus lots of ideas for where to go next in your own pursuit. (An important addition for your library.)

Nontoxic and Natural —
Debra Lynn Dodd

It's all here: 1200 brand names rated for nontoxicity; 400 inexpensive, do-it-yourself formulas; 500 mail order sources.

Diet for a New America —
John Robbins

Don't pick up this book unless you are prepared to change your life: a quick reading will suggest it, and a thoughtful one will demand it.

Alters of Unhewn Stone —
Wes Jackson

Presents a case for a new, prairie-based, sustainable agriculture and the changes required to achieve it.

The Dream of the Earth —
Thomas Berry

Calling himself a "geologian," the author argues for an earth-centered cosmology instead of a human-centered one.

The Rights of Nature: A History of Environmental Ethics —
Roderick Nash

A great companion to his earlier work, this one places deep ecology within Western intellectual history and explains the role of Muir, Leopold, Carson, and others in fashioning American environmental thinking.

The End of Nature —
Bill McKibben

Touted as the successor to Carson's Silent Spring and Shell's Fate of the Earth, this one makes the long overdue case that it is our lifestyles that are destroying the earth. (A chilling read.)

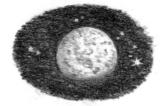

PERIODICALS

"Green Letter" —

A quarterly newspaper for and about the evolving international Green movement (published in conjunction with "Greener Times," the official newsletter of the Green Committees of Correspondence.)

"Environmental Ethics" —

A quarterly journal which seeks to bring together nonprofessional environmental philosophers with the newly emerging professional interest in the subject (but remains a bit heavy for the layman).

"Green Teacher" —

A bi-monthly publication that aims to relate the latest green movement debates to ideas and practice in education for the environment (based at the Alternative Technology Centre in Britain).

"Orion Nature Quarterly" —

A magazine that aims to characterize conceptually and practically our connections and responsibilities to the earth.

"The Trumpeter" —

A quarterly journal of ecosophy (ecological harmony and wisdom) dedicated to exploring a new consciousness and sensibility in our environmental relationships.

"Holistic Education Review" —

A quarterly magazine that explores how education can encourage the fullest possible development of human potentials and planetary consciousness.

"Utne Reader" —

A bimonthly journal of the best of the alternative press. (Don't miss this one.)

"Pollen" —

A new journal for bioregional education designed to promote communication in between the North American Bioregional Congresses.

"The Animals' Agenda" —

A monthly magazine offering a broad range of information and materials dealing with the animal rights movement.

"World-Watch" —

A bimonthly magazine of the Worldwatch Institute devoted to raising public awareness about environmental threats and providing information for those who want to do something about them. (Lots of good data.)

"The Deep Ecologist" —

A multifaceted, seasonal newsletter of the deep ecology network in Australia.

"Mother Earth News" —

A bimonthly publication that now calls itself "the original country magazine," but remains chuck-full of "how-to" articles nonetheless.

"Tranet" —

A quarterly newsletter (and directory) for people who are adopting alternative technologies and transforming their lives.

"Buzzworm, The Environmental Journal" —

A new bi-monthly report on the condition of worldwide environmental conservation, this one aims to provoke (thus its name, an old western term for the rattlesnake). We'll have to wait and see if it lives up to its warning.

"Resurgence" —

A British journal promoting ecological wisdom and a sustainable way of life (with inputs from the philosophical to the practical).

"The New Catalyst" —

A quarterly newspaper covering the alternative movement in western regions of North America (based in rural British Columbia).

"New Options" —

A monthly newsletter that provides a forum for new political ideas and strategies (with lots of reader input).

"Whole Earth Review" —

A quarterly publication that after 20 years still provides challenging ideas, great reviews, useful tools, and other good things.

"Catalyst" —

A periodic newsletter on socially responsible investing and the new economics movement.

"The Ecologist" —

Another British journal that consistently serves up hard-hitting analyses and fresh perspectives on a range of international concerns. Jonathan Porritt called this one, "the doyenne of green magazines."

"Earth Island Journal" —

A quarterly publication that will keep you in touch with what's happening at the forefront of worldwide environmental action.

"Firmament" —

A quarterly publication of the North American Conference on Christianity and Ecology.

"Garbage" —

A "Practical Journal for the Environment" and a mainstream rather than an interest group voice for environmental concerns, this bi-monthly is too new to determine if it will represent a cornucopian response, but do take a look.

"E, The Environmental Magazine"

A brand new publication that promises to be a clearinghouse for the environmental movement and a resource for those who want to improve upon their relationship with the natural world.

Since some of these recommendations are oriented towards the North American scene, if you would like a list of titles and sources for books and periodicals available in your part of the world, send a self-addressed, stamped envelope — or enclose the appropriate international postal response coupons — to the nearest office of The Institute for Earth Education (at the address noted on the insert in this book), and ask for our current earth educator's bibliography.

Finally, if you decide to move farther afield in helping others search for a new lifestyle, you would be wise to shy away from most present-day rural farming areas, and look for one of the new intentional neighborhoods or communities instead. Why? Because farming has changed a lot in the past few decades, and it is not all the farmers' fault either.

Have you been to a farm recently yourself? As an interpretive consultant, I visited a modern farm in Minnesota a few years ago. It was a square mile of flat, unobstructed land. There were no trees, no creeks or ponds, no animals, no cow pies. There wasn't even a barn, just a corrugated steel canopy under which sat several hundred thousand dollars worth of these gargantuan machines. Their tires alone towered above me.

In Family Farming—A New Economic Vision, Marty Strange analyses the rise of agribusiness and proposes a four point mandate for farm policy that could serve as an important step in resurrecting the family farm of yesteryear:

1. A farmer should be able to pay for farmland by farming it well;
2. A farmer should have to farm it well;
3. A farmer should have to pay for land by farming it, and by no other means;
4. There should be no motive for owning farmland other than to make a living by farming it well.

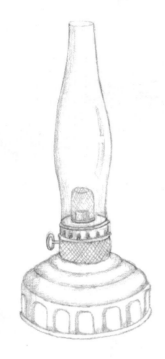

Shortly after my arrival, a fellow clad in a blue jump suit came out of a small office building, climbed up into the cab of a huge combine, adjusted the air conditioning and stereo, and prepared to rumble back and forth across the land harvesting his crop. Before he departed he explained that they had sprayed the soybeans with defoliants a couple of weeks back so they wouldn't be so "trashy." Welcome to agribusiness in the 80's and 90's.

In the western world the family farm of our dreams is a vanishing species, and our collective desire for energy intense lifestyles explains much of the reason why. The family farm was an important unit in our societies as long as there was a reason for the people living there to stay together. For generations that reason was dictated by necessity. Every member of the extended family served as an integral part of the whole, and each person had a role, an important job to perform. But with the advent of fossil fuels the energy became available for all of that to change. Using wood and horses as energy sources, a farm family a hundred years ago enjoyed a fairly direct, personal, familiar, stable relationship with the land and its life. Today, that way of life has vanished. Using large quantities of fossil energy instead, farmers now have a fairly indirect, impersonal, unfamiliar, unstable relationship with the earth. Meanwhile, their kids are likely off in some sports program and the grandparents have moved to the sunbelt.

In one of his novels Wallace Stegner relates the story of a farm family living on the Great Plains in Canada during the 1800's. Each evening they would light the kerosene lamp and gather around the kitchen table in a regular family routine. Father would likely be looking through seed catalogs, mother would be working on a piece of sewing, and the kids would be busy with their lessons or hobbies. All of that changed, however, when father came home one day with a marvelous new kind of kerosene lamp that cast a lot more light than its predecessor. Enjoying its brightness the members of the family began to spread farther out into the room each evening — eventually leaving mom at the table alone. Gone was the closeknit feeling of that evening ritual where everyone clustered around the table together. Fortunately, this mother proposed a new rule: even though the light energy was available to move farther away from one another, they would

CONFLICTING DREAMS...

Don't you love those armchair farmers in the cities who talk these days about family farms and the values inherent in them, then go out and cast their ballots for the corporatization of their country? Too many of us still have "Little House on the Prairie" in our hearts and stock prices in our heads.

not do so. Everyone would still gather around the table each evening just as before.

That was one perceptive mom, ahead of her time. She grasped the risks that a new technology always brings, and decided that her family would control the technology instead of letting it control them. (Ironically, in our present century I understand that some of our modern, busy churches have found it necessary to set aside one evening each week for family night at home.)

Doesn't it make you angry that many of the people who talk the most about simple values in our Age of Excess end up supporting those who aspire primarily to collecting material goods? It appears that a lot of people want their cake and want to jump on it too. Can we really have simple values and live complex lives?

As I mentioned in the beginning of the book, current polls indicate that the great majority of people believe that "science" will still find the answers to our problems. As if the "answers" reside somewhere else, and science will eventually deliver them to us on a perpetual December 25th. Clearly, we have failed to get our message across to people. Science has already found the answer, and it is simple: we must produce and consume less, and enjoy more.

"PARTICIPATING IN ENVIRONMENTAL PLANNING AND ACTION"

In education much of the environmental action in the past has been a case of too much, too soon. We stirred the kids up and set them off to do battle with the forces of evil before they knew enough to realize it wasn't "us" and "them" — it was all of us together.

Consequently, lots of kids today know that pollution is a problem; very few of them realize it is their fault. As I explained earlier, in earth education we do not believe it will do us much good in the long run to have the kids deal with a local environmental issue unless they deal with themselves at the same time. Saving local natural communities is a laudable objective, for example, but we have to be careful that it doesn't just

The key to a good life is not having what you want, but wanting what you have.

A NEW CATALYST...

One of the projects we have on the drawing board in The Institute for Earth Education is to help people start "Harmony Houses" in their own communities. These would be regular, suburban homes that earth educators turn into more environmentally-sound dwellings, centered upon more environmentally-sound lifestyles, then invite their neighbors to come in and take a look at what they are doing. Contact us if you would like to be involved.

SKILLS FOR ACTION...

Of course, the skills for effecting change are also a necessary concomitant of the driving thrust of environmental concern, particularly at the adult level. And even though one would assume young people are already picking up such abilities in their regular social studies programs, if they are not and cannot, then we are going to have to provide them. However, in earth education, such skills need to be seen as a necessary outgrowth and expression of personal environmental commitment, rather than as tools of power to effect change for its own sake.

serve to siphon off the kids' concern and energy, concern and energy that might better be applied to their own lifestyle changes. Saving wild areas should be the natural end of all our other work, not both the beginning and end of our environmental commitment.

While it may be true that we become the roles we play, what we want people to become are deep ecologists, not transient environmentalists. We want them to do more than merely participate in one local issue as kids, then fade away as adults. If we are going to get them to "stay the course," we need to get them working on the one issue that they can actually do the most with over time — their own lifestyles. There is a nice payoff here educationally: it is also the one place where they can be guaranteed success.

In our chapter on Acclimatization we noted that one of our principles has always been that heightened feelings combined with increased understandings form the matrix out of which positive environmental action arises. Without the undergirding of that matrix we believe many of the attempts at environmental action will result in only temporary "world-making" activity on the part of most of the adolescents involved. There is no doubt that they will enjoy it, and they will probably accomplish some good things, but in the end, we must ask if they got wrapped up in it for the wrong reasons.

Once again, I am not arguing against environmental action projects for young people. Far from it. I am just questioning from where they arise, to what they are applied, and when they are instituted. We believe the key is, after your participants have acquired some basic understandings and feelings about the earth and its life, to get them to focus on what they personally do in their own day-to-day lives. Not what their parents do, but exactly what do they do — with their time, their money, their belongings, their space, their requests of others, etc. In short, we don't want them to go home and lay a rap off on their folks; we want them to lay a rap off on themselves.

Assuming the kids have already been working on their own environmental habits for a while, why not ask them to come up with ideas for how they can work together on ways of making it easier for all of them to become more environmentally sound? Challenge them to figure out new ways of doing things in their collective use of energy and materials, and support them in their efforts to refine and implement those changes in their surroundings.

Why not sponsor an annual "Living Lightly" Conference where the students can conduct sessions on new ways of doing things personally and getting things done collectively? Invite youngsters a few years older than they are to come in as your keynote speakers and share the changes they have made in their lives or the changes they initiated in their schools and neighborhoods. Kids know that their teachers go off to conferences to learn new things so why not have them do the same. (And be sure to use all the trappings and tools of an adult gathering.) Try to maintain a balance though between sessions on how individuals went about doing something different and how groups worked together in environmental planning and action. Perhaps you could join with other schools in your area and rotate the hosting function each year from school to school.

Frankly, if we are going to tell the kids to "think globally, act locally," then we had better do a bit of it ourselves as adults. That is why have added a phrase to René Dubos' line. In earth education, we say, "Think globally, act locally, be personally."

It has been said that in the end we want people to become environmentally literate citizens, but in earth education it is more than that. We want them to become active participants in the process of desired environmental change. As we have seen, kids can start where they are (at school and home), while their parents can begin at their own level (at the same time). What we really want the adults to do is to participate in local environmental efforts (and to support national and international ones). We want them to get out there and get started in their own neighborhoods. After all, once they have a handle on changing their own lifestyles, it is only natural that they should begin working with others to develop the systems necessary to support their personal efforts. What I am talking about here though is new *systems* for maintaining positive environmental effects over time, not band-aids for patching up a problem.

When I first came to George Williams College in the seventies, I noticed that each year someone would

DO AS I DO...

There is a delightful story about a mother bringing her son to Gandhi and asking him to tell the boy not to eat sugar. But Gandhi merely told them to come back in two weeks. When they returned, the mother asked Gandhi again to please tell her son not to eat sugar. This time Gandhi simply said to the youngster "stop eating sugar." In exasperation, the mother asked Gandhi why he hadn't done that on their first visit, and he replied, "Two weeks ago I was eating sugar."

try to get a recycling project going there. Cardboard boxes with attached recycling signs would appear in the hallways, and then gradually disappear as they filled with other refuse. The whole effort would usually last until either the initial zeal wore off for its sponsors or until the maintenance folks decided it represented too much of a mess. And the next year would see the whole sequence recycle itself. In fact, that appeared to be the only long-term recycling accomplished.

Consequently, when I became the advisor to the Student Environmental Alliance, my advice on recycling was *not* to do it. That's right. I was convinced that what was needed was not another recycling project, but an institutionalized *system* for maintaining the recycling of aluminum over time. I urged the students to spend their energy designing and monitoring such a system (and educating the campus about its use) and let others actually implement it.

Although not much has been happening in the environmental movement on the education side, lots of good things have been happening on the *environmental action* scene for some time (habitat preservation, recycling systems, toxic waste control, etc.), and the *environmental studies* people continue to contribute much to our understanding of the problems (pollution data, species extinction, ozone depletion, etc.). As a result, there are lots of places to go for ideas on how you can work with other people in your own area to implement systems for carrying out environmental improvements. We have listed several environmental organizations in the accompanying box. Everyone involved in the work of earth education should belong to at least a couple of them. Or, as a participant at a conference I spoke at suggested, we should all be giving at least 1% of our annual income to various alternative environmental causes.

Environmental education emphasizes output; earth education focuses on input.

RECYCLE
↓ HERE

aLTERNATIVE eNVIRONMENTAL ORGANIZATIONS

Planet Drum —

Fosters bioregional development and communication. Publishes the biannual review, "Raise the Stakes."

Greenpeace —

Leads the international environmental action movement. Publishes the bimonthly "Greenpeace Magazine."

Rainforest Information Centre —

Supports campaigns to save the world's rainforests (from its base in Australia). Publishes the periodic "World Rainforest Report."

Earth First! —

Bills itself as a forum for the no-compromise wing of the environmental movement. Publishes eight annual issues of "Earth First!", the radical environmental journal.

Regeneration Network —

Facilitates interaction, communication, and information exchange on regenerating "communities" of people who are improving their lives and living environments. Publishes the bimonthly "Regeneration Newsletter" and maintains the Regeneration Network Database in the U.S.

Environmental Action —

Serves as a national environmental lobbying and educational organization in the U.S. Publishes the bimonthly "Environmental Action Magazine."

Earth Island Institute —

Aims to act as an environmental early warning system, identifying the issues needing immediate action around the world (formed by David Brower, former director of the Sierra Club and founder of Friends of the Earth). Publishes the quarterly "Earth Island Journal."

Green Committees of Correspondence —

Reports on the work of numerous grassroots Green organizations in the U.S. Publishes the newsletter "Greener Times" and a journal, "Green Synthesis."

Eleventh Commandment Fellowship —

Supports the ideal of a Christian deep ecology based on an additional commandment. Publishes "The Eleventh Commandment Newsletter."

Co-op America —

Offers an alternative marketplace for socially responsible individuals and businesses, and provides people with practical strategies for integrating their economic choices with their politics, lifestyles, and values. Publishes the quarterly, "Building Economic Alternatives" and an annual catalog of the products of cooperative workplaces.

Friends of the Earth —

Serves as a global, action oriented network of environmental advocates. (Although the American branch may no longer fit this category, check out the other international groups). Publishes the bimonthly newspaper, "Not Man Apart."

For a list of the names and addresses of other alternative organizations in your part of the world, send a self-addressed, stamped envelope — or enclose the appropriate international postal response coupons — to the nearest office of The Institute for Earth Education (at the address noted on the insert in this book), and ask for a roster of our recommended organizations.

GUIDES FOR ACTION...

Our Earth, Ourselves, Ruth Caplan

The Green Consumer, John Elkington, Julia Hailes, and Joel Makower

Saving the Earth: A Citizens Guide to Environmental Action, Will Steger and Jon Bowermaster

However, when it comes to actual participation in most local environmental actions, helping people overcome their inertia may be our primary task. Even as leaders, we sometimes think the problems are so large that we are powerless to really do very much about them. Even though we intellectualize that we can really make a difference individually, a numbing lethargy seems to grip all of us when faced with the complex array of environmental problems that need our urgent attention. In addition, many of our fellow role models who were concerned about the environment in the sixties have already succumbed to the euphoria of the eighties. For them, getting ahead has replaced getting in touch.

There is a positive note in all of this. Over the years all the pollsters have reported that there is an amazing, deep-seated concern about the environment in the United States. Nothing quite like it has surfaced before so consistently and pervasively over and over again in the polls. There is no doubt, people are genuinely concerned, and I suspect it is the same in other countries.

The problem, of course, is that because most people really understand so little about how life functions here they can be easily misled, and we all know there are lots of folks out there playing the Pied Piper role. But just think what would happen if we could connect up educationally with the interest that already exists. If we could tap into that environmental concern as educators, build a solid base of feeling and understanding under it, and help people start working on their own lifestyles as they assimilate those feelings and understandings. . . . Woweee! We could see some very rapid change out there, and it would be none too soon.

Obviously, we cannot depend on most of our politicians nor even many of the leaders of environmental education to help us out with our task. We are going to have to do it ourselves. In the U.S., in the 1850's, the political leaders of a midwestern state, responding to a bill introduced in their legislature that called for the protection of the passenger pigeon, said: "The passenger pigeon needs no protection. Wonderfully prolific, having the vast forests of the North as its breeding grounds . . . no ordinary destruction can lessen them, or be missed for the myriads that are yearly produced." So much for the results of counting on our own leaders. Duane Elgin expressed it this way in his book on alternative lifestyles:

> *"Just as we tend to wait for our problems to solve themselves, so too do we tend to wait for our traditional institutions and leaders to provide us with guidance as to what we should do. Yet our leaders are bogged down, trying to cope with our faltering institutions. They are so enmeshed in crisis management that they have no time to exercise genuinely creative leadership. We may keep waiting for someone else. The message of this book is that there is no one else. You are it. We are it."*

"The whole theory of the universe is directed unerringly to one single individual—namely to You."

— Walt Whitman

Maybe if we just focus on individually reaching a couple of people, it will make it easier to get started. Someone set up an experiment a while back to see how difficult it would be for an individual in New York to personally get a message delivered to someone she didn't know in Los Angeles. The ground rules were that the person in New York could only telephone someone she knew with the message, and that person could only telephone someone he knew, and that person could only telephone someone she knew, etc., until the final person said, "Oh, yeah. I know her. She lives down the street. I'll deliver the message." The astounding result of this experiment was that it only took about five phone calls to get the message delivered. Why? Because each of us knows several hundred other people. You can just picture how it would have worked: "Let's see, Martha. Didn't your brother's kid move out to California? Right. We'll call him." "Gosh, hello Martha and Bill. Somebody in Long Beach? Yeah. I think one of the secretaries lives out that way. I'll call her." "Hi, Greg. Sure, my boyfriend's family lives over in that area. . . ." It's the old pyramiding technique from sales, but it works. If each person can just reach a couple of others, and each of them can reach a couple of others, and each of them can reach a couple of others, we can together change the world. So let's get started.

Internalizing (understandings). Enhancing (feelings). Crafting (lifestyles). Planning (actions). These four components make up the last WHAT of earth education — the *Processing* that must take place if we are to succeed in helping people live more lightly on the earth. We want these processes to become a regular, integral part of our participants' lives, for they deal with the very essence of the environmental change required for a healthier planet. However, once again, we believe you will discover that they become synergistic as you work with them, blending together in a way that makes them more than the sum of their parts.

Okay. We have examined the three WHATS of earth education — the Understanding, the Feeling, the Processing. Granted, it is a lot to digest, but perhaps thinking of them as "the head, the heart, and the hands" will help you hold onto them. Just remember, earth education set out to do what environmental education appeared to want to do but didn't. So we must not make the same mistakes ourselves. The point of earth education is change. If there is no change, there is no point. In the next chapter we will look at the three WAYS earth education uses to achieve that change: *Structuring, Immersing, Relating.*

TEN GREEN VALUES...

Adopted at the founding meeting for the Committees of Correspondence (an American network of grassroots green organizations), these values characterize much of the green movement around the world, and provide a good starting point for environmental planning and action on a larger scale.

1. Ecological Wisdom
2. Grassroots Democracy
3. Personal and Social Responsibility
4. Nonviolence
5. Decentralization
6. Community-based Economics
7. Postpatriarchal Values
8. Respect for Diversity
9. Global Responsibility
10. Future Focus

CHAPTER fIVE

eARTH eDUCATION… tHE WAYS

eARTH eDUCATION...
tHE WAYS

STRUCTURING

*"We believe in building complete programs
with adventuresome, magical learning experiences
that focus on specific outcomes."*

STRUCTURING
WAYS

Structuring is our way of talking about the skeletal
framework of earth education programs. Of course,
when we use that term, we are talking about building
complete programs that include all three of our WHATS
(the understandings, feelings, and processings). We
are talking about how we design our learning experiences
for each of those components in a focused, sequential,
cumulative fashion. And since we want our model
programs to serve as powerful springboard experiences
for what takes place back at school and home, we are
also talking about how we make those activities inter-
active and dynamic. Finally, we are talking about how
we put those pieces together into a whole — an overall
program that is so carefully crafted and so rich in
detail that it becomes synergistic, or more than the
sum of its parts.

Since the last section of this book will deal with
program building in some detail, our task here is to
look at some "structural" guidelines that we follow in
both crafting and conducting earth education experiences. We have broken them down into four things to
do and four things to avoid.

Things to Avoid: naming and labeling
                talking without a focal point
                playing twenty questions
                drifting into activity entropy

Things to Do: create magical learning adventures
                focus on sharing and doing
                emphasize the 3 R's:
                      reward, reinforce, relate
                model positive environmental behaviors

"aVOID NAMING AND lABELING"

Now before you get your stones out, let me explain a couple of things about this one. We are not saying that we're opposed to identification in all situations. In other words, it is not that we think names are unimportant across the board, but that they are relatively unimportant in earth education, and I think that's saying two different things.

Years ago, I told my Acclimatization staff that I did not want them to get all caught up in that endless parade of the names of things, that tired, old procession that still seems to captivate so many leaders in nature education. Instead, I recommended that we deal with broad categories of things as much as possible in our work — a tree is a tree, a bird is a bird, a rock is a rock, and so forth.

However, it wasn't long before they came back saying, "Now, Steve. Isn't it all right for us to name something just once in a while, when it's really important?" They sort of had me at that point, and sadly, I replied, "Yes." As it turned out that's how I discovered that most staff members in this field arrive on the scene with what I call Pandora's Identification Box under their arms. Inside those containers are the hallmarks of the educated child in our societies: the names of about 20 trees, 20 wildflowers and 20 insects. It did not take long either to find out they were like Pandora's Box in mythology, for the staff couldn't open them just a crack and reach in and pull out a name just once in a while when it was really important. Instead, as soon as they opened the lid, out poured the names of 20 trees, 20 wildflowers and 20 insects. In fact, the only way I ever got a handle on my staff about this problem was by telling them that if they opened Pandora's Box out there, they would die. Later, I made believers out of them.

You see, I just didn't want them to get bogged down in all those names of things. We believe the concepts are more important than the names. We want to focus on the processes of life instead of its pieces.

Nonetheless, even if you are into names, here are a few general problems that you should consider. To

begin, I think it was Alfred North Whitehead who said it best: "When you name something, you tend to stop thinking about it." Consider your own experience for a moment. How often have you walked down a trail with a group when someone asked the leader, "What's that?" And the leader replied with something like, "One-sided Pyrola," while the group continued right on walking? That was the end of it. People had a name for it and they stopped thinking about it.

I remember reading some medical research years ago which indicated that if you gave someone with a psychosomatic sort of illness a name for his malady, it would tend to go away. (Presumably, because after receiving a name for it, the person would stop thinking about it, and it would disappear.) Anyway, I found this mildly interesting, but didn't think much more about it until I returned from one of my annual journeys to the everglades. One winter after getting back from Florida, I came down with a pain in the right side of my chest, and it seemed to be getting worse by the hour. Finally, with much reluctance, I decided to see a physician. You have to understand that since I had not been to a doctor in well over a decade, that was a major decision for me. Anyway, I went to one of those private clinics where they run you through a battery of tests all day, poking and probing at you like a side of beef. At the end of that ordeal I found myself sitting in the doctor's office in my underwear — trying to look cool. When he finally arrived, he hit me on the shoulder with a filing folder of my test results, saying, "Hey, you're in pretty good shape." I replied, a bit impatiently, "Wait a minute, Doc. What's this pain all about in the right side of my chest? That's what I'm in here for." "Well, it's a lot more common than you probably think. Show me where you hold the steering wheel of your car." At this point he was quickly reinforcing some of the reasons I had not been to a physician in well over a decade, but I finally answered, "Okay, like this I guess." "I see here in your file that you just returned from a trip to Florida. I bet there was a lot of traffic on the interstate highways after the holidays, hey? In fact, I bet you probably drove almost straight through without stopping." I replied, "Yes, as a matter of fact, I did, but what's that got to do with this pain in my chest?" "Well, you've just strained the muscles across your chest from all that tension and stress as you fought the steering wheel all those miles. It's called inner chest cavity syndrome." At this point,

"I am ashamed to think how easily we capitulate to badges and names, to large societies and dead institutions."

— Ralph Waldo Emerson

I started laughing and couldn't stop. "It's a syndrome, huh? I think I'm being punished, Doc." Needless to say he thought he had another problem on his hands (I couldn't stop laughing at the time to explain the crotchline reaction story from Acclimatization), but wouldn't you know it, my pain seemed to go away that same evening. I had a name for it and thus stopped thinking about it. I could go home and tell everyone, "No problem. It's just inner chest cavity syndrome."

By this time, someone is probably reading all this and saying to himself, "Now, Steve, all that's fine, but you can't really love something and cherish it and respect it unless you can name it." I don't think that's true; I love lots of things I can't name. I love the mosses and lichens and mushrooms, for example, but I don't really have names for very many of them. What most people mean when they talk about naming like this, appears to be laying off on others those particular names they happen to have accumulated for a handful of the pieces of life. In earth education we believe that approach is like sharing your collection of travel slides; for most people, it gets boring rather quickly.

Or how about the whales? Gosh, I love the whales, but I don't really have the names of very many of them worked out. Oh, I have a lot of whale names floating around up in my head, but if you brought a batch of whales in and lined them up in front of me, I doubt that I could name very many of them. That doesn't mean though that I don't love them, or cherish them, or respect them. It just means that I haven't had much need to hold onto their names. If you are into whales and need to know all their names for your research, that's great. You are the one that should really learn all those names. We just don't think you should try to turn most people onto whales (or anything else) by dumping all your names on them.

One of the challenges that I have often heard over the years in our workshops has been, "Well, you name people don't you? Just look at these name tags you make us wear. So why can't we name other things?" I am afraid that's confusing two different items. In fact, if we named the trees like we name people, I might be for it. In that case, we would have to come up with individual names for each tree, like Sally Sycamore, Susie Sycamore, Sam Sycamore, etc., and learn them based upon individual characteristics, just like we do

"They who know the truth are not equal to those who love it, and they who love it are not equal to those who delight in it."

— Confucius

SALLY SYCAMORE

with people. You would really have to get to know those individual trees to be able to distinguish Sally from Susie from Sam. Wouldn't it be fun to sneak into an arboretum at night and rip down all those traditional signs, replacing them with individual names like these? That sure would freak the staff out when they showed up for work the next morning.

For most folks, probably the most persuasive argument in favor of naming has been the one about the name being like the tab on a filing folder, i.e., once you have the filing folder labeled it is much easier to add new information you find later. The problem here is that most people spend most of their time getting these huge, elaborate filing systems ready, but then never do anymore with them. Remember learning all those names for your leaf collection, and insect collection, and rock collection? What else did you ever put in all those filing folders? If you became a botanist, an ento- mologist, or a geologist, that's fine. At least somebody got some use out of that idea. But overall, I am convinced that that approach has produced the greatest number of empty filing folders in history. Besides, our whole point in earth education is that we want to build filing folders for the processes and places and positions in life instead of all its parts. We will come back to this analogy again in the next chapter, but for now, we believe most nature educators spend their time setting up the wrong kinds of filing systems.

What about having the kids name things based on their own observations? This is perhaps the easiest trap to fall into. We did ourselves. I think the key here is whether or not the kids really want a name for themselves, or whether the leader just uses this device to work them around to a name the leader wants to lay off on them. As far as I can recall, in all the years we conducted the original Acclimatization program, there were only two things for which the kids really wanted names. And in both instances, they simply named those things themselves. It was easy and logical, and it grew naturally out of the activity. Our experience has been that when people of any age really want a name for something, they will come up with one. So why foist a name off on them when there is no need?

A variation on this game of trying to work the kids around to making an observation that will "reveal" something's name is the tack some leaders take in telling

Actually, over the years
I have softened somewhat
on my insistence here,
explaining that broad
categories could include the
name of a whole kind of
animals, like whales, or
could describe a niche, like
shorebirds, but I still don't
see much need for going
beyond that point in most
situations. Besides, most of
your participants are going
to arrive with a few common
names in hand. That's okay,
just don't get upset if they
are not the same ones you
have collected for your own
files.

the kids they are just learning the names of some new
"natural friends." However, you usually pick your own
friends in life instead of having them chosen for you by
someone else. And what about the danger of forming
"natural cliques" this way instead, i.e., where you only
know a few things and thus tend to focus almost
exclusively on them. In fact, studies indicate that in
such cases your eye tends to focus only on those things
you have names for and thus you miss all the rest.
That is a high price to pay I think for knowing the
names of a few of your leader's friends.

I can almost hear somebody out there reading
these examples about now and crying out, "What
about poison ivy, Van Matre? What would you do
about that?" Folks, poison ivy actually makes my case
for me. It is a classic example of where people know
the name, but not the thing. How many times have you
heard people exclaim, "Poison Ivy. Oh, no! Where?"
And they're standing in it! Most people have learned
the name given to this plant, but they haven't learned
anything about *it*. In fact, that's how we handle those
kinds of problems. In the case of poison ivy, we say,
"Folks, we'd like for you to get to know this plant because its sap causes a rash for some people." What is
important in this case is that the participants get to
know this particular kind of plant. They check out its
size and shape and color, its texture and pattern. They
note where and how it lives; who its neighbors are and
what kind of community they find it in. Most of all,
we try to convey the attitude that there is nothing
wrong with poison ivy, and they can call it anything
they want; what is important is that they are more
familiar with it than its name. As with everything
else, its name is really the least important thing to
know about it.

During the piloting of our original six-hour
Acclimatization program, we had a professor of botany
on our staff. She wasn't involved in the actual piloting,
she was doing some other things for us, but I thought
it would be good for her to see what was going on with
the ACC effort. So I asked her to tag along with a
group each week and see what they were doing. About
halfway through the summer, she came to me and said,
"I guess I'll stay." I asked what was wrong, and she
replied in a shaky, albeit exasperated voice, "Steve,
you act like you just don't understand how hard it's
been on me. All summer long those kids have been
walking by those hemlocks and calling them pines."

I apologized for my lack of sensitivity and said, "Hey, Martha, you've been tagging along with the ACC groups for half the summer now. Is it really important, *in ACC*, that the kids have the word hemlock for that particular kind of tree instead of the word pine? You tell me. If you think it is really important, we'll design a new activity out of which will come the word hemlock."

My approach here was that I didn't want the leaders to be constantly correcting the kids. Our axiom has always been: we rebuild, not correct. So I said that, if necessary, we would create a new activity to do the job of rebuilding. Anyway, about a week later we met again, and I asked, "What have you decided, Martha? Is it really important in ACC that the kids have the word hemlock instead of pine?" Frankly, it was a tough question for a professor of botany, but she finally answered in a mumbled whisper, "Well, maybe . . . I don't know, I guess not — maybe it really isn't that important." It was a big hurdle for her.

Please don't misunderstand what I am saying here. It's fine for professors of botany to have all those names. That is exactly who we should expect to learn them. We just shouldn't turn them loose on the public.

"Okay, how do you guys answer 'What is it?' questions? Aren't you stunting intellectual growth if you don't give them an answer?" No, I don't think so. That is just another one of those justifications for laying off all your names on everyone else. First of all, you should explain to your participants before the experience why you are *not* going to be naming things. Once they know the purpose of an earth education program, most people, young or old, understand our aversion to traditional labeling quite easily. However, keep in mind that the "What is it?" question may still pop out from time to time. People often use this question when they are curious about something. It's shorthand for, "Hey, look!" Based on many years of experience, we developed our own checklist for the staff to follow in answering that query. . . .

"You must not know too much, or be too precise or scientific about birds and trees and flowers and watercraft; a certain free margin and even vagueness—perhaps ignorance, credulity—helps your enjoyment of these things."

— Walt Whitman

aNSWERING
"WHAT iS iT?" qUESTIONS

☞

☐ Use action. Take a closer look yourself. Examine it. Use several senses. Ooohh and ahhh, a bit. Most of the time you will find people were just curious about it. (Don't be afraid either to just grunt in response. That's a humorous way of getting people out of the traditional pattern.)

☐ Use silence: don't say anything at first. Always count to ten very slowly before commenting. Often that's when they will remember that you are not going to be naming everything. (But don't just stand there, follow #1 above.)

☐ Use a one line description of its process (system) or place (community) or position (niche) instead of its name. (It's a decomposer, a bird of prey, a flying insect-eating mammal, a prairie flower, etc.)

☐ "I don't know what the experts call it, but if it's really important for you, maybe we can figure out some things about it, and come up with a name ourselves." (Just don't use this to maneuver someone around to your label.)

☐ "If you really need a specific name for it, I can help you find one later. But meanwhile, let me suggest that you look at it differently. . . ."

☐ "Well, remember we're not going to have time to do a lot of naming, but if it's really vital for you, some people around here call it. . . ." (This should only be used as a last resort and be sure to qualify it so the participant will still see the name in a regional context.)

Finally, I like to tell new earth education leaders, "Names are like landmarks; you just don't need very many of them to find your way." Think about it. What if you stopped someone downtown to ask for directions, and they began identifying each and every store for you, block after block, all the way to your destination? It probably wouldn't take long before you would say, "Hey, wait a minute, just give me a couple of landmarks and I'll find my way." Well, the same holds true for things in the natural world. And since most people will have already picked up a number of these landmark names on their own, you won't have to worry about introducing very many of them. However, the above guidelines notwithstanding, if an individual youngster appears to genuinely need a name, consider whispering it to her. Maybe she needs that landmark on her road of life.

Names are like landmarks; you don't need very many of them to find your way.

I know it's true. Some people do find a measure of comfortability in being able to name things. Perhaps it is their way of taking stock of their surroundings. (They certainly seem quite satisfied with any old name you pluck out of the air for them. Try it and see.) However, even though they fix things in time and space by naming them, they usually don't relate them to one another and thus remain forever ignorant of the more dynamic dimension of life. Some of this may just be a need to have power over things, but I suspect for most folks it provides some sense of security for them. It may even be a sign of the immaturity of their relationship with the earth; if so, we need to help them get beyond that stage.

Let's be honest though, it is the traditional nature educators who usually have the most trouble with this guideline. Those who are into "N & N's" (names and numbers) are often not very friendly towards us. It's like we have denied them their identity, i.e., being able to name everything in sight. When we come along and ask them to refrain from getting their strokes in this way, they naturally feel threatened.

I remember one professor at a workshop years ago who got up during the question and answer session to ask me, "What's wrong with these kids knowing some *facts*?" I should have realized I was in trouble when he turned to the audience to present this question, and before I could respond, he launched into a lengthy monologue on the subject. We had just finished a session that afternoon dealing with our Sunship Earth program, and I had listed on a flipchart the seven basic ecological concepts that Sunship Earth is based upon. So as he continued his harangue about the importance of facts, I moved slowly over to the flipchart and put my arm around it (thinking I could call attention in this way to those understandings). However, my non-verbal commentary was lost on him, and I finally had to explain that there was nothing at all wrong with the kids knowing some facts and that these were the facts we wanted them to know.

It was only in a conversation with him later that I realized that for this particular leader the "facts" were the names and numbers, not the processes. Enumeration, not revelation, was his game. He actually still believed that you had to start out with all the names of things (he called them building blocks), then work your way up to the communities and systems. I was almost speechless. In the face of the overwhelming evidence in our societies that this hasn't worked, he still clung tenaciously to "the way he had been taught."

Obviously, this is going to be an extremely difficult change to make in this field. Some people do build good feelings about the natural world via learning lots and lots of names (and they often go on to become the educational leaders of the next generation because there's little else they can do with their collections), but this approach just hasn't worked in mass education. It appears that a rather small percentage of people get really interested in nature in this way. We are simply missing the majority of folks out there. In fact, instead of turning them on as intended, we have often turned them off.

Okay. We are aware that this is a tough guideline to follow. It's a hard one for us, too. In fact, if you really insist that your staff try to follow it, you are probably going to wipe out about half of everything the average leader in this field has to say and do. As you will see later though, that is actually an important reason for the guideline, because it will force everyone to do something else.

"A new truth does not triumph by convincing its opponents and making them see the light, but rather because its opponents eventually die out, and a new generation grows up that is familiar with it."

— Max Planck

"avoid talking without a focal point"

Piaget said that you learn with something in your hand. Why? Because it's in the concrete for you. As much as possible we have always tried to have something in every learner's hand in our activities. However, we realized early on that in mass education that was not always practical, but if we couldn't provide something for every learner to hold onto, we were sure that at least we could have something for every learner to focus upon — a visual focal point. You see, from the beginning I didn't want our staff standing around talking without something to help them make their subject more concrete, especially since ACC was supposed to involve lots of firsthand contact. I wanted something specifically set in every learner's visual field, and we have always gone to great lengths to get that job done.

I remember watching a leader one day talking with his group at the beginning of our hour-long "Day in the Bog." Usually the group would start out on the beach, cross this small isthmus of woods to the bog, then crawl around the bog on their hands and knees before recrossing the wooded area and returning to the spot where they started from on the beach. On this particular occasion I was watching the leader from my customary perch in a tree, but I couldn't see any focal point down there at all. The kids were sitting in a circle on the sand and the leader was rattling on and on at great length about something obviously important primarily to him . . . yackety, yackety, yackety. Meanwhile, as you can imagine, the kids were busy digging holes in the sand, tossing it at one another, etc. After the activity was over, I got hold of the leader and asked, "Bill, what in the devil was going on down there? I didn't see any focal point at all, and you must have been talking nonstop for fifteen or twenty minutes." Bill replied with a comment I will always remember, "Well, you know Van Matre, you have a lot of bright ideas, but sometimes they just don't work." After pantomiming the removal of that knife, saying, "Here, you may need this again," I asked him to explain. He said he had been trying to tell the kids how the glaciers formed the bog. How massive amounts of ice had built up on the land in that area and when the ice began to melt back a huge chunk of it had broken off at that spot. As this large piece of ice melted, the rocks and gravel that it had scoured up as it moved across the land crumbled out around its edges to form a small ridge of debris. Then as the ice continued to melt, this ridge got larger and larger until in the end you had this small pot lake formed inside the rock and gravel outwash from the glacier. Finally, this trapped body of water had grown over with a mat of plants to form our bog. Bill concluded his explanation with, "What do you expect me to do, get a piece of a glacier for a focal point?" I replied, "Great idea, Bill. Let's get a chunk of the glacier." So we froze up a chunk of ice with sand and gravel all through it, pulled it out of a gunny sack at the beginning of the activity and plopped it down in the middle of the circle of kids on the beach. It was just heavy enough to form a depression in the sand, and as it started melting in the sunlight, the rocks and stones inside began crumbling out around the edges to form a small ridge. This way most of the leader's words could also be eliminated because the real punch was going to come at the end of the hour when the kids returned from crawling around in the bog. That chunk of the "glacier" not only made a great focal point, it made a great demonstration for what had actually taken place there as well.

As you can see, we go to great lengths to establish focal points. In fact, if you really insist that your staff follow this guideline, you are probably going to wipe out about 50% of everything the average leader in this field has to say and do. (That's a little test to see if you are still with me, for in the first guideline we wiped out about half and in this one about 50%. It's true, just with those two guidelines much of the repertoire of many outdoor leaders will be eliminated.)

In short, we believe traditional outdoor leaders often have a compelling tendency to talk too much, to rattle on and on in too much detail, to stray away from the point, to stroke themselves instead of their learners. In our work, deciding upon what not to say is often more important than choosing what to say. We believe a few, carefully selected words spoken in association with an appropriate item of focus — a natural object, a prop, a symbol — will do more for the learning process in most cases than a lecture. At best, listening is still too passive — in earth education, we are after more *action*.

FOCUS CAREFULLY...

When following this guideline, be sure that you don't make yourself or a previous activity the focal point. We want the learners to remember the message more than its delivery. Focal points are immediate and concrete things that highlight and clarify and demonstrate the intended message.

"AVOID PLAYING TWENTY QUESTIONS"

Our third guideline deals with our questioning strategy. If you have stuck with me so far, you have probably figured out that we are going to have something weird on this one, too. We do. In fact, it is probably the strangest guideline of all.

You see, from the beginning I did not want our staff to rely on the most common educational technique ever devised. I did not want them to play Twenty Questions all the time with their learners. Oh, you may not know it by that name, but I am sure you know the game; that is the one where you start out with a one or two word answer in your head to a question the kids didn't ask, then work them around to your response. It is played every day throughout our societies in hundreds of variations. "Kids, does anybody know . . . blah, blah, blah?" "Kids, why do you think . . . blah, blah, blah?" "Kids, what is the . . . blah, blah, blah?"

I am convinced this is a very dangerous educational game in our field. Why? Let's look at an example of how the game is usually played. The leader asks the group a question, then begins working their responses around to his answer. "No, not quite. Remember what I said about. . . ." "Well, that's close, but not really it." "Hmmm . . . that's a different idea. Anyone else?" "No, I'm afraid you're off track." "Okay. Hooray! That's it, exactly." Sound familiar? It should. We have probably all been in many situations like this ourselves. But what is really going on here? This leader just managed to get several people to process incorrect responses in order to get one person to process a correct one. You see, when you say something out loud you set up a stronger mental pattern than when you think it, or when you just hear someone else say it. That's why oral drill can be helpful. That is why verbalization is important. The learners are doing something with the information. So the leader in this case has helped more people establish stronger mental patterns for an incorrect answer than for a correct one. Wouldn't it seem logical, if you were going to play this game all the time, that at least you should go back and ask those who gave the incorrect response to repeat the correct? In fact, you should probably have them say it

"The only interesting answers are those which destroy the questions."

— Susan Sontag

twice — once to offset that they said the incorrect and another time to build the correct. Of course, no one does this; it would take all day. But that is exactly why I suspect we ended up playing Twenty Questions so much to begin with, it is a substitute for good doing. It is less time consuming, so we think we are being efficient. However, that's not true if lots of folks are either learning incorrect responses or not playing the game at all.

Watch almost any group of learners (of any age, in any setting) who are being subjected to the Twenty Questions approach. Who is really playing? I suspect that most of the time you will find a few bright, competitive, articulate, over-achievers playing along, while everyone else listens impassively, or takes on the glazed look of stuffed animals. Why? Because lots of kids learn at an early age that the secret in these situations is not to risk giving a wrong answer. They learn a slightly different version of Lincoln's famous saying about remaining silent ("It's better to remain silent and thought a fool than to speak up and remove all doubt"). Kids learn, it is better to remain silent and thought a fool *by the teacher* than to speak up and remove all doubt *in the eyes of your peers*. After all, in silence there is safety. So a lot of kids just don't play the Twenty Questions game.

Who does play? Remember that old adage, the best way to learn something is to teach it? Well, no wonder, for who is really doing most of the doing in these situations? Right, the teachers are. In fact, I like to tell leaders fond of playing Twenty Questions all the time that they should really pay their participants for helping them learn all that stuff. Can't you just picture a leader going from kid to kid passing out money at the end of an experience and saying, "This is for you, and this is for you, and this is for you. . . . Thanks, and come back tomorrow." Kids wouldn't miss school would they? "We have to go, 'cause they pay us to help the teachers learn."

It is also rather ironic that the kids who need the help the least (the highly verbal) are often the ones doing most of the responding in this game. They are having a heyday while the rest of their classmates are left feeling bored or lost or stupid. Today, playing Twenty Questions with a group of any age immediately sets the leader up in an unfavorable role for many of the participants. Since many people associate this approach with past feelings of inadequacy, they simply withdraw. They see the leader as a teacher-tester instead of a learner-helper. They feel like they are back in school.

Okay, if Twenty Questions is so awful, why do leaders play it so much? Because that is what they have seen done and then did themselves. Everyone tends to do what they see, and learn what they do, not just the kids.

To question or not to question, that is the question.

I believe Twenty Questions is actually a degraded, now damaging form of the socratic method. Have you ever looked at the original conversations this idea is based upon (the socratic dialogues)? Let me tell you, in reality, you rarely get a situation like that. (Don't feel too bad. Historians are not sure Socrates ever did either.) Those guys were all perfect. They asked exactly the right question at precisely the right time. And no one ever came right off the wall with a response. Besides, Socrates had some mind, hey? How many people do you know can really sit down and carry on a carefully structured, directed conversation like that (let alone with 30 kids)? I mean, I can't even remember what I was thinking about in the shower this morning. Is it really very practical to think that lots of leaders can play the role of Socrates? Where are they trained to do this? The truth is, they are not. Check out the college catalogs and see how many courses you find in socratic dialoging. Consequently, most leaders just sort of stumble along doing what they have seen done. When you are digging into the original dialogues, note the kinds of questions Socrates was asking too. You won't find very many with clearcut answers. He wasn't fooling around with fill-in-the-blank stuff, but things like the nature of beauty, good and evil, fun things like that.

Anyway, the early educators said that that was a good way to teach because it taught students how to think. So now leaders use it to have kids figure out the name of a tree. Poor old Socrates, we have mostly forgotten what he had to say about beauty and good and evil, but we are still quizzing the kids to try to bring about learning.

No, I am not saying a leader should never use the so-called socratic method. Given the right objectives, a reasonable setting and a few ready learners, it can be dynamite. However, for much of mass education its use is limited, and particularly for education out-of-doors that is supposed to have a large dose of firsthand contact and active doing.

In the end, what leaders are doing with this technique much of the time today is playing a sort of word game — with verbal clues — that bears little resemblance to the original socratic dialogues anyway. It is more like they are leading their participants through a mystery about what is in their heads instead of what is in the kids' heads. Think about it: they use Twenty Questions as a self-assessment tool when working with a group of kids, and if someone parrots back the answer wanted, they feel successful (*I* taught them). However, if the students don't get it right later on the exam, *they* failed — they didn't learn. It's an ironic twist on the original intent, I think.

If that is not enough, there are other problems as well. The first of these is what I call the impulsive responders. Apparently, a certain percentage of people in these situations tend to respond not to the complete question presented, but with something that one word in the question triggered in their minds. We are not sure exactly how this happens, but it is probably through some form of mental association. Consequently, some people answer questions with something totally unrelated to the point. (Frankly, they are also people who don't seem to have any incubation period between the arising of a thought and its delivery.) I am sure you have seen this happen; they just pop out with the first thing that pops into their minds, and it is usually a bit embarrassing for everyone. In adult groups, you often see perplexed leaders try to act like those responses didn't actually occur, and the people standing next to the impulsive responders look the other way too. It's like we all agree that if we don't look at them, they won't be there. Unfortunately, kids are not so under-standing and usually laugh loudly at their unfortunate classmates.

Another problem is that a few good players tend to dominate the action. They also tend to manipulate the leaders whenever possible. To be honest, I was one of those players myself. It is another illustration of how what you really learn in school is what you actually spend much of your time doing there. I can still remember being asked by my classmates at the beginning of a period to "get him off on something." For, like many students, I had learned that what you do in school is ask the teacher a question before he can ask you one. If you were good back in those days, you could get him off on war stories for the whole period. (I wonder how the kids get by today since not as many teachers have war stories to tell anymore.)

Unfortunately, I learned a lot of this myself after Acclimatization had been underway for a while. (If you look closely at our first book, I am afraid you will find too many examples of the Twenty Questions technique in there as well.) Whenever I would observe our staff leading ACC activities, I would note that quizzing the kids in that way represented a pretty low level form of doing. As a result, I pulled the staff together one day and told them that if they asked any questions from then on, they should only toss them out to the whole group without putting anyone on the spot. Furthermore, I suggested it would be a good idea if they put the questions off on themselves, like, "Gee, I wonder if . . . ?"

That strategy sounded pretty good, but it didn't work either. For a long time I couldn't figure out why not, but you could just sense that something was wrong in the way the kids were responding. And do you know what it was? The leaders were tossing their questions out to the whole group, but they were still expecting answers, and the kids were picking up on those nonverbal expectations. The kids thought it was just the usual Twenty Questions game in a slightly different form.

Anyway, I brought the staff in again and told them I had added a new clause to the guideline: "From now on, ask your questions rhetorically, but don't expect answers." That's right, *don't* expect answers. Well, if you are like my staff, I have probably pushed you about as far as I can at this point. They said something like, "What in the devil is wrong with you? You start out by telling us we can't identify things, after we just learned the names of all this stuff in our courses at the university. Next, you say we can't talk unless we have something to focus upon, but we are *intellectual* aren't we? Finally, you say we can ask questions, but shouldn't expect answers. But why in the world do we ask it if we don't want an answer?" They were definitely riled up. And I replied, "Don't expect answers, expect action."

DON'T FALL OFF...

Sorry, but I think this is the great abyss in nature education. All too often what passes for nature study today is the leader taking the kids out and playing follow-me-gather-round, inter-changing, if you will, Show and Tell with Twenty Questions. The kids don't _do_ very much at all. Here we have them outside the classroom at long last, but our techniques are still classroom techniques. It doesn't make much sense. Didn't we always justify outdoor learning by saying that we had to take the kids outside in order to do those things we couldn't do inside? Well, in reality, what passes for outdoor learning in a lot of cases is the same thing people were doing indoors before. The only difference is that the learners now stand around playing Twenty Questions instead of silling around.

In our work, we ask our questions to get better doing. We ask questions to facilitate the flow of the activities, not to quiz the kids.

For example, if you had a group of kids in a circle around an old tree stump and asked, "Gee, I wonder how it feels?", you would be asking your question not to get an answer, but to get the kids more involved with the stump. You would be asking your question to get more contact and better doing.

Frankly, this may be the most difficult one of our guidelines to follow. It has been for us. For example, during the development of our Sunship Earth program we would have major design sessions in the midwest, then send the activity descriptions off to our pilot site in Oregon. I remember getting one of those sheets back one day with a new recommendation written on the margin. . . .

In this particular activity the kids were all seated on the ground and each of them had a large handful of soil to explore. The "new" idea was for the leader to ask, "Hey kids, how long do you think it takes to make a handful of soil like this?" Of course, I went right through the roof when I read this. Why is it that we so often play Twenty Questions with things kids don't know anyway? I got the leaders in Oregon on the phone and said, "I bet this is what happens. You ask that question and someone responds, 'I don't know, man — a thousand years.' And you say, 'No, not quite that long,' then start working them 'til you get someone to say '200 years.' The point is, if it's really important for the kids to know that it takes a couple of hundred years to make a handful of soil, why didn't you share that with them? You knew, didn't you? They didn't know, so why did you ask them? I mean, couldn't you just say, 'Hey, kids, you know it takes a couple of hundred years to make a handful of soil like this.' Wouldn't that be much quicker? After all, the soil is concrete for them right then, and they are doing something with it, so why not just share with them something you know?"

Later, after my initial outburst, I called back again and said I still wasn't sure about what was going on in that activity. I asked if it would be all right with them if the kids went away thinking that it took 278 years to make soil? Wouldn't that be close enough? Okay, how about 124 years, would that be satisfactory? We're not going to quibble over a few years here are we? So what would they think about 72 years? Finally, I got an exasperated, "Steve, it's important for kids to know that it takes a long, long, long time to make soil!" And I replied, "Then how come you are doing it? If it's so important, why doesn't the activity do it? I thought that was the activity's job."

This time the response was something like, "Okay, bright boy, what would you do?" After thinking about it, I said, "Hey, if what we want the kids to know is that it takes a long, long, long time to make soil, then it's simple. Let's have them try and make some." And that's what we did. After they had explored their handful of soil, we gave them cloth bags and wooden mallets and asked them to try and make some soil. They would gather materials from the forest floor, then just beat the tar out of those bags. They would dip them in the stream and pound on them with rocks. They would team up and jump up and down on them with their feet. Let me tell you, they put a lot of energy into trying to make some soil, and after each attempt, they would open their bags, check the contents, and sigh, "No, not yet." Guess what they learned very quickly? "Man, it takes a long, long, long time to make soil." And that was the point.

USING QUESTIONS: PROBLEMS AND SOLUTIONS

First, there are four main problems with the Twenty Questions approach:

☐ It is not active enough. It represents a poor substitute for good doing in outdoor learning.

☐ Most of the time there are more incorrect responses being processed than correct ones so we are strengthening the wrong things.

☐ Most of the doing is done by the leader and a few articulate participants. Lots of people don't play, and thus don't do anything.

☐ The impulsive responders tend to disrupt the flow of the activity, while the good players tend to manipulate it for their own purposes.

Questions, however, are not always bad. Here are some ways in which questioning can facilitate good outdoor learning:

☐ The leader asks questions of herself that do not require a response but guide exploration: "I wonder what causes it to look like that?" "I wonder what's underneath?"

☐ The leader asks questions that result in concrete action: "What does it feel like when you rub it up against your cheek?" "Can anyone find a smooth stone around here?"

☐ The leader asks questions that don't have specific answers: "How did it make you feel?" "Which understanding made the most sense to you?"

☐ The leader asks challenging questions of the whole group: "Can you all find an example of _____?" "Do you think it's possible to trace _____?"

☐ Since we want to change the perception that the leader is a teacher-tester, the leader becomes someone else, such as an inspector or a reporter, and then asks the question in that role. (After all, it's now his job! However, the leader should over-play such a role just a bit to remove any potential threat.) Please note: this technique should only be used to clarify and reinforce after an activity has set up the understanding.

Once again, eliminating most of the questioning places the responsibility for the learning on the activity and not on quizzing techniques for bringing out the understandings. Make up tasks to perform or mysteries to solve instead of questions to answer. (And don't forget the importance of motivation and reward.)

Okay. I know I am bludgeoning you with this one, but I think playing Twenty Question games represents one of the major problems today in the entire field of outdoor learning. Here are four additional examples of how it can backfire:

ALLIGATORS AND SHOES

In one of the original wild animal parks in the U.S. the leader begins her morning in an auditorium full of energetic fourth graders by asking, "Kids, kids, does anybody know what alligators are good for?"

Remember the problem of the impulsive responder? Out of nowhere comes the response, "Shoes!" The leader jumps in quickly with, "Oh, no, no . . .", but she never does get them back on track. For the kids the whole thing was a game, "Yaha! We got her!" And to this day some of them probably believe alligators are good for shoes; after all, they heard it at the zoo.

tEPEES aND aNTELOPES

In a new museum, highly praised for the designer's attempt to make the artifacts feel more accessible, the leader still gets the kids inside, sits them down, and, you guessed it, plays Twenty Questions with them. . . .

Picture a group of kids sitting on the floor right in front of an authentic Plains Indian tepee. It has all the customary paraphernalia around it, inside and out.

(Observing situations like this I often play a little game with my graduate students called the "Guess-what-the-leader-is-gonna-pick-up-and-ask-them-what-it-is?" game. My scholars get very good at it. On this occasion, the leader goes over and picks up a small set of black horns. . . .)

"Kids, does anybody know what these are?" This time there was no problem with an impulsive responder, for the leader couldn't get any response at all. "C'mon, kids. Doesn't anybody know?" Frankly, these kids didn't have a clue that he was holding up a set of antelope horns. After all, they were sitting in the middle of a large metropolitan area miles and miles from any antelope. In fact, I thought the best response was the kid who finally called out, "It's a baby buffalo, man!" I would have probably bought that myself. Why not? I thought that was pretty good for a little set of black horns. But the leader was obviously frustrated. He was hot on the trail of "antelope," and seemed about ready to give up when he noticed a little girl right on the front row waving her arm frantically. Actually, she had been trying to get his attention for some time, but he was standing so close to the group that he had been focusing mainly on the back rows. And since the kids on the front row didn't want to sit with their heads bent back at 45 degree angles, they were largely ignoring him.

Anyway, the leader seemed greatly relieved to have

someone volunteering a response at last, but instead of telling him what he wanted to hear, the little girl asked, in an excited voice, "What's that?", pointing to a small hole in the side of the tepee right at her eye level. However, instead of doing something with this, the poor leader, who was locked into trying to get "antelope" out of someone, snapped back in frustration, "Oh, it's just a mistake!"

The little girl withered visibly (they always know when they've been put down), and the leader went on with his game. Think about what he could have done with that incident: "Wow! I don't know what that is. Look here, kids. Do you suppose it could be a bullet hole? Or maybe an arrow went through here. Let's see if one will fit." Wouldn't the kids always remember that one? Can't you see them bringing their parents and friends back to see where the arrow went through the tepee? Or maybe the leader could have even replied, "I don't know. Let's see, maybe it was made by an *antelope* horn."

NUMBERS AND ZOOS

At a large metropolitan zoo the leader has the kids who have just arrived gather up on the carpet in one corner of their meeting room, and begins their visit with the question, "Kids, how many animals do you think we have here in our zoo?" One little girl responds with 100. The leader smiles, shakes his head, and calls on the student next to her. But instead of going for a much larger number this girl says 200. This time the leader says, "No, more than that, lots more," and waves his hand to indicate a larger amount, but it doesn't work. When the next kid responds, I know the leader is in trouble. She merely continues the pattern that has been set up by the previous answers and tries 300. At that point, I lean over to my host and whisper, "If the next little girl says 400, he's in a lot of trouble." You see, the kids didn't know what he was talking about anyway. They thought it was a counting exercise. You know, 100 - 200 - 300, and the leader couldn't get them to make the jump to a higher number (and he has 4000 animals in his zoo!). Sure enough, the next girl in the row said 400, and I got to laughing so hard I was doubled up. Next came 500, 600, and so on, and they had to get me out the back door. It was going to take all morning just to get an answer for the opening question.

"Grown-ups love figures. When you tell them you have made a new friend, they never ask you any questions about essential matters...Instead, they demand: 'How old is he? How many brothers has he? How much does he weigh? How much money does his father make?' Only from those figures do they think they have learnt anything about him."

— Antoine de Saint Exupéry
The Little Prince

Since you might be saying to yourself at this point, "Okay, smart aleck, what would you have done?" I should explain that I don't really know. It depends on what I was trying to accomplish. I can't figure out why it was important for those kids to know how many animals they had there, but if that was my goal, if that was what I really wanted the kids to take home with them at the end of their visit, I think I would have had them count them. Every one of them. Wouldn't they remember that one all their lives? "Hey, let me tell you, we spent one whole day counting all the animals out there and they've got a lot of them suckers. It was awesome!"

MAMMOTHS AND WHALES

In a well-known children's museum, highly touted for all of its hands-on material, the teacher brings the kids in and sits them down right in front of the skeleton of a baby wooly mammoth hanging from the ceiling. (Picture several large, open shelving units arranged in a half circle all around the group, with an amazing array of weird things scattered about on their shelves. There are huge beetles from the jungle, large fossil bones, chunks of rock specimens, strange stuff all squished up in big jars, etc.) The leader comes in rather nonchalantly, walks up and sits down in front of the group, right underneath the skeleton of the baby mammoth.

First of all folks, no leader in his right mind would compete with the skeleton of a baby wooly mammoth. Not only that, but by sitting down in front he practically guarantees that a lot of the kids won't see him. Next, he proceeds to pull an old cardboard box out from under a bench and slowly removes a folded up poster from it. Never mind that every kid today knows that nothing of any importance ever arrives anymore in an old, dilapidated cardboard box, because the kids aren't paying any attention to him anyway. As you can imagine, they are entranced by all the amazing things around them: "Ohhh . . . Frankie, look at that thing over there in that jar!"

Anyway, the leader unfolds his poster and holding it up, asks, "Kids, does anybody know what this is?" (It's a picture of a dolphin jumping out of the ocean. At the time, I think to myself, you have to be kidding.

This is taking place in one of our planet's largest cities. These kids — the few whose attention he can gain — won't have a clue.) "Hey, kids. Take a look up here. Doesn't anybody have an idea what this is?" The last thing I heard as I darted out the door, growling in frustration, was: "It's a whale, man!" But I doubt that he bought that. . . .

Finally, let me wrap up this guideline with one last warning. There is one word in particular that you should be on the lookout for in our field. It hides a lot of bad stuff in education. In fact, it almost always comes with its own punctuation mark, and it most often follows weak educational experiences. To be honest, every time you see it, you should be suspicious of whatever precedes it. Chances are good you will find that it is used most often in an attempt to get a weak activity to do something it didn't do. The word: discuss (colon).

Some leaders will even admit that their activity did not work because they will say right up front in the description, "You have to pull it out of them." I cannot imagine admitting that your activity failed like that, but people do. And of course, in many instances the word discuss really turns out to be merely an educational euphemism for playing that old game again — right, you guessed it — Twenty Questions.

"avoid drifting into activity entropy"

In natural systems the term entropy is used to describe the tendency of all energy to degrade towards increasing disorder and practical uselessness. The same tendency holds true for activity systems. Good learning experiences usually require focused and sustained inputs of energy, particularly in mass education where one leader tries to work with lots of learners. However, the natural inclination of many leaders is to put less and less energy into an activity over time.

The whole thing usually starts out innocently enough, and at first, what's going on is so deceptive that no one notices. For example, take the activity we call Micro-Trails. In this experience the participants use a 30' piece of string to lay out the route of their "micro-trails" branching off from a central path. Along the way they designate the neat discoveries they make using ten red-topped sticks which serve as station markers. Their final tool is a magnifying lens for closeup viewing of the things they will share when they take each other on a tour.

When activity entropy sets in, it is usually the station markers that get left out first. "Isn't it all right if we get rid of these sticks? They're a real hassle, and the kids always forget to pick them up afterwards. Can't we just have them gather up some small sticks lying around on the ground? You know, that'd be more natural, wouldn't it?" But it's not long before these leaders are back with a request to eliminate the string. "Hey man, can't we do away with this string? I mean it gets all tangled up, you know. Can't we just tell them to make a trail? Why limit their freedom anyway with just thirty feet?" Next, it is the hand lenses that become a hassle. "I don't see why we can't leave these magnifying lenses out. I mean, the whole point is to look closely, isn't it? So why can't we just say, 'Hey kids, look close'?"

In a relatively short time, the energy required for this activity has degraded to where the station markers, the string, and the hand lenses have all been dropped. Guess what, though? It isn't very long before the same leaders return saying, "You know, Van Matre, that Micro-Trails activity really doesn't work very well." No wonder, the leaders just eliminated all those things that helped make the activity a success. Those props helped provide structure, and magic, and purpose. Once you dropped them, the activity lost its focus for the participants. They couldn't visualize what to do.

Unfortunately perhaps, an enthusiastic leader can still pull this activity off occasionally without all those props, but then the leader burns out on it more quickly because it takes so much more personal energy to make it work. So the thing you must watch out for when following this guideline is that normal tendency to seek the easiest route at any given moment. In this case, it is deceptive and seductive.

What happened to our original blindfold walk in the Acclimatization program provides another good example of activity entropy in action. We first used the blindfold there during our "Day in the Forest." We had a marvelous section of climax forest at camp that

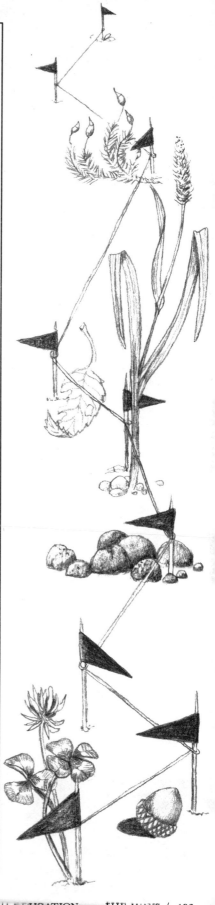

I wanted the kids to experience in a very powerful way. It was a small, closed canopy forest, with huge old trees, some so large that three kids could not hold hands and encircle them.

We started out on our "immersion" in an adjacent area of forest that was made up primarily of smaller fast-growing, sun-loving trees. It was an area that had been clear-cut not long before. (In such areas in northern Wisconsin the lumber companies are fond of putting up signs along the roads that read, "Aspen Regeneration.")

First, we had the kids lie down on the forest floor like spokes in an old wagon wheel. With all of their shoulders touching, they could look up at the sky and the leader could talk very softly about it and still be heard clearly. We asked them to check out what colors they could see overhead, and they replied with mostly blues and whites as they could see the sky and clouds easily through the open canopy of those small-leaved trees. Next, we asked them to dig their hands down into the layer of leaves and stuff beneath them to see how it felt, and they responded that it felt mostly dry and sandy.

Afterwards, we got them up and spaced them out along a rope that we had knotted about every four feet. We explained that we were going to blindfold them, but they shouldn't worry for as long as they held onto the rope everything would be fine. We went to great lengths to explain that they shouldn't be afraid of bumping into something either as we walked along, because we had checked the route out very carefully. They should just hold onto the rope and absorb all the sounds and smells and feelings of this special experience we had prepared for them.

After putting large blindfolds on everyone, we started out very slowly walking in silence through this rather brushy, open-canopy, sunlit forest towards the climax forest they had never seen. As soon as they entered the edge of the climax forest, the differences began hitting them rapidly. First, the sunlight disappeared, and then the temperature dropped. Wherever they were going, they knew it was darker and cooler. Then the sounds began to change as well. Not only was it less crunchy underfoot, but the noise of their own movement seemed to be sucked up and away from them. As they walked deeper into the area, their feet

told them it was becoming softer and spongier too. And it smelled moister and richer. Wherever they were, they knew it was a lot different from where they had been.

Finally, we asked them to take off their blindfolds and lie down again like spokes in a wheel. This time when we asked what colors they could see overhead, we got mostly yellows and greens as the sunlight filtered down through the closed-canopy high above. When we asked them what the forest floor felt like here, we got words like wetter and deeper and earthier.

The point is the kids had experienced in a rather dramatic way some of the differences between these two kinds of forest communities. That is why we had used the blindfolds in the first place — to intensify those differences. Sadly, when we published our book outlining this activity, lots of people seemed to focus on the blindfold as a technique and missed seeing the real purpose of the experience.

Several years after the release of <u>Acclimatization</u>, I was sitting in a departmental meeting one morning on the Wisconsin campus of George Williams College where I had just arrived to assume a teaching position. Being rather bored, I was looking out the window of the meeting place when I spotted a grad student leading a group of kids up the asphalt path outside the cottage. The participants were holding hands, had their eyes closed, and were literally dragging one another along. The last little girl in line was getting sort of whiplashed by all this pulling and tugging, but she was keeping her eyes tightly shut nonetheless. Unfortunately though, she was suddenly whipped head-on with a resounding smack into a tree right next to the path (actually, the only one in the vicinity). I jumped up and rushed out fully expecting that she would be split open like an old pumpkin. Instead, she just had a big knot on her forehead, but I grabbed hold of the leader, pulled him aside angrily, and asked, "Bill, what in the devil do you think you're doing?" His reply startled me. In a shaky voice, he said, "Hey, I'm just doing your thing, aren't I?"

I knew right then that we were in a lot of trouble. Activity entropy had set in with a vengeance. And in this case, the leader had not only eliminated the rope and the blindfolds, he had done away with the forest as well.

All this doesn't mean that we think you should never improve upon an activity. There are always new and better ways of doing something. But it does mean that you should be very careful about making changes. You should ask yourself over and over whether your change is really for the participants or for yourself. That's the real danger with activity entropy; it sneaks up on you and convinces you that you are actually doing something good for the learners.

So how do you avoid drifting into this disordered state of affairs? First, pay careful attention to the details of any activity. Earth education activities are designed to maximize learning, and all the parts of each activity are important: the atmosphere, the props, the touches, the sequence, the flow. Often a tiny detail plays a vital role in the activity, be it in connecting one part to another, motivating the learners, or providing a convincing setup.

Second, when preparing for any activity, you must analyze every aspect once again, visualize every detail once more. Even if it's your hundredth time, you must examine your site, arrange your props, rehearse your role, check your sequencing. And you must do it all with the same care and attention to detail as if you were just starting out.

Finally, although the activities themselves are the most important teacher, it is the leader's energy that sparks the action and provides a catalyst for the built in elements of magic and adventure to do their job. We tell our leaders, "When you're in doubt about an activity, put some energy back into it."

By energy we don't mean just more action either. It is also the effort you put into the preparation, the dramatic tone you add to your voice, the way you use your hands and body. It is the adventure created by the look in your eyes. It is the time you take to put a hand on a shoulder, give a pat on the back, place a note in a bottle. And it never means uncontrolled rowdiness either. The energy put into an activity is always directed toward a particular outcome.

Frankly, earth education is not an undertaking for someone looking for the easy way out. There is none. But that's not to say that the hard work involved is not enjoyable. One of the surprising realizations

GETTING YOUR JUST REWARD...

Besides making the learning happen with greater ease, paying attention to such details often results in a "bonus" for the leader. The learners are usually very much aware of the care and concern a leader puts into an activity and this often results in an increase both in their interest in what they are doing, and in their respect for the person leading them. Our old adage, "If you care, they'll care," has never let us down.

about our work for many people is that our activities are as enjoyable for the leader as for the learners. Plus, there is a built-in quality control factor — when you do something you enjoy, it will usually be done well.

All right. We know these are not easy guidelines to follow. Basically, what we are asking you to do is to shift your entire focus on our field. It is almost like you will have to reach up there in your head and change the perceptual lens through which you have been seeing this kind of work. It's tough, but we are convinced it holds the best promise for building and leading successful programs in living more lightly on the earth.

Now that we have set some parameters for what you should not do in earth education, let's turn our attention to some important things you should do.

"dO CREATE MAGICAL lEARNING aDVENTURES"

Learning about the earth and its systems of life should be a joyous experience. If we force someone into doing it, we run the risk of diminishing the very understandings and feelings for which we strive. Instead, we must entice our learners into our activities. We

Sadly, in our fast-paced, energy-intense societies today a lot of people will not let themselves relax enough to feel this kind of emotional attraction. Kit Williams, the author of that marvelous children's book for grownups, <u>Masquerade,</u> submitted his work unsuccessfully to over a dozen American publishers before finding one willing to make his magical story available here. The other publishers all thought no one would be interested in solving the mystery of where Kit buried a real treasure in England. But his simple romantic tale, combined with his colorful and intricate pictures containing the clues, proved fascinating to thousands of Americans as well.

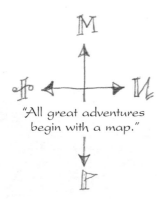

"All great adventures begin with a map."

must pull them along with us and get them wrapped up in the doing. There are two indispensable elements for helping us accomplish this: magic and adventure.

We believe magic is an irresistible influence for young and old alike, yet this mysteriously enchanting element of our work almost defies description. Suffice it to say that magic promises something that will dazzle and entrance you, even though it may be offered quietly and softly. Whatever the setting, magic personally grabs onto you and tugs. Don't worry, you will know it when you feel it, for a surge of emotion wells up inside you in response to its pull.

What we are talking about here is educational magic, not stage magic. Stage magic creates amazement through the unexpected and unexplained. Twenty-two clowns emerging from a small car defies our preconceived notions about reality. In fact, over time people actually seem to look forward to such tricks. They are quite willing to suspend belief and be swept along by unfolding events. Why? Subconsciously perhaps, I think it is because they want to repeat those feelings of wonder and awe that they had as children. They agree to buy in, anticipating that their reward will be a return to the freshness of experience they enjoyed in their youth.

The magic in earth education involves some of those same characteristics that stage magic has always depended upon, but here we go the magicians one better: we invite our participants to join the show. They don't have to watch it; they can become a part of it. Of course, they may have to suspend belief a bit in some situations just as they did for the stage magician, but we hold out a special promise for them. If they are willing to accept a pinch of fantasy to set our stage with us, then they can become actors instead of audience. We say, "Step right up, folks. And *join* the show!"

That's where the adventure comes in. It's the show itself. Only in this case, since you are going to be a part of the cast, you have to get ready for your role beforehand. And getting ready is an important part of any adventure, for an adventure doesn't usually happen in the classroom or at home, that is where you get ready for it. As we all know, adventure happens "out there."

Of course, the very essence of adventure is immersing yourself in new sensory stimuli. On an adventure you know you will see and smell and hear and taste and touch new things. That is why adventures in the natural world work so well; you always have the continual bubbling and churning of life to stimulate you.

Almost by definition an adventure also means a change of scenery. It means going somewhere else (or maybe doing something out of the ordinary nearby). You usually have to undertake some fairly elaborate preparations (maps, guidebooks, films, etc.) before you depart. You make lists, learn new skills, study new words. . . . And an adventure almost always involves special items like food, clothing, containers, and so forth. You have to get things ready to take along (and think about what you might bring back). From daypacks and overnighters to picnic hampers and steamer trunks, these physical preparations are all-important in building anticipation for what's to come.

Educational adventure then involves some of the same steps that you would undertake for any similar experience in your life. You plan and prepare. You anticipate and set forth. You seek excitement and overcome adversity.

GETTING THERE IS BEING THERE...

At heart, an adventure is a break in your routine. In fact, little adventures can be had almost every day simply by changing some small part of your usual ritual as you go from home to school or office or centre. Try it and see.

You're never alone out there

When you think about magic and adventure it helps to focus upon two key words: caring and daring. In

"Magic involves caring; adventure includes daring."

PLUS ARDUOUS PREPARATION...

After devoting an entire morning to getting ready for a brief activity, one of our trainers explained our "magic" this way, "It's three people spending three hours setting up for a twenty- three minute experience." Like all important things in life, your sense of magic will be as strong as your commitment to it.

our Earth Caretakers program, for example, the magic is in the special package delivered to the classroom, the accompanying notes from the slightly mysterious rangers, and the riddles they ask the learners to solve. The adventure comes in the preparations for the day away from school, in the discoveries of the rangers' tools, and the activities themselves in a natural setting.

By its very nature magic also takes forethought and attention to detail, but in this case, it is the leader, not the participant, who has to get ready. Remember the story of Richard in Disney World (from chapter two)? Even though what that father did was a simple act with wonderful results, it required planning and preparation. He visually projected himself and his son standing along the sidewalk at the parade. He saw what was needed to make it a uniquely magical experience for Richard, and he got his "props" ready to achieve it.

Don't be fooled on this one. Magic is not easy. It is not just an extra dose of energy or a dollop of exhibitionism. It is caring deeply (like Richard's father) about the experiences of others. It is both watchfulness and playfulness in adding that extra element.

Our key concept statement for magic could be a paraphrase of the line from the greeting card folks: "When you care enough to *give* your very best." And caring includes taking. Sometimes "giving" *your* very best means "taking" *their* very best. Remember Bill's treehouse? Bill was a good taker. He let the kids act out their fantasy, their dream. He didn't get in the way. He didn't turn it into his dream.

In earth education we have to be very careful to keep the kids on center stage, and limit ourselves to supporting roles. This is hard, especially for some outdoor leaders who may have got involved in the first place seeking starring roles for themselves. You will have to keep reminding them: the only stars in earth education are the learners.

Just as magic has an element of adventure in it, adventure has a bit of magic. You always go adventuring in the belief that something special will happen. You expect to discover new things. It is that delicious sense of setting out just knowing that you are going to find something new that propels you over the hurdles. And you are willing to take some risks in order to make those discoveries happen. Adventure then always includes a bit of the unknown. That is what we mean by the daring. After all, you have to be willing to leave the audience in the first place and join the show. You have to take a chance.

MAKING MAGIC

- ⊕ Weave a story
- ⊕ Add a pinch of fantasy
- ⊕ Set up a discovery
- ⊕ Work on the details
- ⊕ Prepare a surprise
- ⊕ Send a secret message
- ⊕ Confound with something amazing
- ⊕ Suggest a presence
- ⊕ Use clues and riddles
- ⊕ Create an appealing atmosphere

- ⊕ Do the unexpected
- ⊕ Add something to make it unusual
- ⊕ Forecast events to build anticipation
- ⊕ Consider all the senses
- ⊕ Wear a costume
- ⊕ Watch for special moments
- ⊕ Demonstrate your care with light hearted gifts
- ⊕ Remember to be a good taker

When you go on an adventure you also expect to deal with the unexpected — changes in the weather, mechanical failures, etc. They become almost a predictable occurrence for the daring journey you have undertaken. They don't even seem as troublesome as they would at home. They are now part of the action. It reminds me of that old saying in education, "you learn when the adrenalin flows," which is probably why we remember the worst experiences in such situations as well as the best ones.

The success of any adventure has a lot to do with how you viewed it in the beginning. In this case, perception seems to be governed largely by anticipation. (In fact, someone has said that perception is merely

"An adventure is an inconvenience rightly considered."

— Chesterton

the difference between what you see and what you expected to see.) Depending upon your expectation the same basic activities may be seen as exciting adventures or boring tasks. It is the difference between *assigning* a group of kids to hike to the end of the lake, and *challenging* them to undertake a perilous journey to record their names in the ledger buried there. How you present any activity will have a lot to do with how someone responds to it.

adding adventure

setting it up

⊕ Offer a challenge

⊕ Suggest a trip

⊕ Reverse the commonplace

⊕ Send a map

⊕ Leave a note in a bottle

⊕ Introduce a mission

⊕ Ask for help

⊕ Present a mystery to be solved

⊕ Organize an expedition

⊕ Pretend it's a bit risky

⊕ Add an unknown element

⊕ Include a break in the routine

⊕ Create vivid images of the possibilities

⊕ Bring in something exotic as a focal point

⊕ Make it multisensory

getting ready

⊕ Gather data (guides, maps, photos)

⊕ Chart your course

⊕ Study the terrain

⊕ Review the bare necessities (food, water, shelter)

⊕ Prepare your gear

⊕ Practice needed skills

⊕ Learn about the natives (language, customs, costumes)

⊕ Decide on your role

⊕ Make a journal

⊕ Think about what you're going to bring back

In the end, magic is something that is usually arranged for you by others, while an adventure is something you play some part in arranging for yourself. True, you can establish the conditions for a magical experience in your own life, but it is difficult because your expectations may then overwhelm the reality. At best, you can put all the elements together, then just wait and see what happens. As the Samurai warriors put it, "Expect nothing, be prepared for anything." Finally, when it comes to the adventure, keep in mind the role of perception. What seems like an adventure for one person may be rather tame for another. That is why we need to include a variety of elements wherever possible in our educational experiences, and provide opportunities for the learners to be involved in the preparations.

Most importantly, please remember that our adventures are first and foremost focused learning experiences. Just as we don't want to let the magic overwhelm the message, we must not let the adventure overcome the aim. In other words, be sure you don't let your means become your ends. Begin your planning with a clear grasp of your educational outcome and make sure your subsequent activities do not blur that objective. This is tough work. You will have to keep coming back, refocusing, and starting in again. Frankly, a lot of leaders go astray at this point. Don't fool yourself. It takes a lot of discipline to keep at it until you have crafted a magical learning adventure that actually does the necessary job.

It may also help to think of the adventure as the assimilating part of your learning model. At least, that way you will always clearly identify your outcomes and set up your learners for them beforehand. But we will look at that model in more detail in our chapter on building your own program.

Finally, an old guard naturalist asked me at a session recently if there wasn't someplace left for just taking youngsters out and sharing with them some of the wonders of the natural world. Sure there is, but the task today requires much more of us than that. We can no longer afford the luxury of taking a few learners (either literally or figuratively) on an occasional nature hike. We have to reach more people with more of a message, and get them started on the lifelong process of forming better environmental habits. To do

MAGIC IS AS MAGIC DOES...

Some people have objected to the use of the term magic because of its supernatural connotations, but something can be "mysteriously enchanting" (as the dictionary defines it) without being supernatural or unearthly. Others have objected on different grounds, saying that nature itself is magical and all our props and costumes and "gimmicks" just get in the way. As we have pointed out repeatedly ourselves, there is some danger that this may occur, but we believe that earth education, properly implemented, enhances and intensifies the magic of nature. That's why we have worked so hard at infusing our experiences with it.

that we have to use some techniques from mass communication and education and that's where the magic and the adventure comes in. There will always be a place for an enthusiastic nature lover to share some of the marvels and mysteries of the earth with a group of learners, but we are fooling ourselves if we believe that that is the answer for dealing with our environmental problems.

"do focus on sharing and doing"

Remember: in our work it is not show and tell, but share and do. I know there is a subtle difference in that line between the idea of show and share, but I think it is an important difference. We don't want our leaders to convey to the learners an attitude that they are going to teach them something; we want them to convey an attitude that they are going to share something with them.

As I noted in the section on the WHYS of earth education, I don't like the word teach very well. In all the things I have written, you will seldom find that term. I think it conveys the idea of a leader acting *upon* people instead of *with* them. And we want leaders who will join their participants in exciting learning adventures.

The real task of teaching is creating exciting learning situations.

When I was an educational consultant, I liked to go into an inservice teachers' training session after school and start out by saying, "Folks, I'd like to ask you to do just one thing for me — please, please don't teach." You can imagine the reaction to this opening as the faculty began pulling back in their chairs, whispering comments to one another: "Oh, great. Where did they get this one?" "Every year they drag somebody in here." "I wonder what they paid him?"

After the mumbling subsided, I would explain that I was quite serious because I was convinced the task of teaching is *not* teaching, that is why it is so hard. The real task of teaching is creating exciting learning situations.

The main difference between show and tell and share and do obviously comes in the telling versus the doing. If we are serious about mass education, then we must accept that what we tell people is not as important as what they do with what we tell them. That is the essential step. If we are really going to help our participants learn something, then we must come up with things for them to do with our information, and we must create an atmosphere of pleasurable sharing in which that doing takes place.

Make no mistake, folks, I am well aware that you are not going to learn the stuff in this book just by reading it. The only reason I can get by with this at all is because this isn't mass education. By the very nature of this material you are pre-selected and pre-motivated learners. (If you have read this far, you are probably not even being made to do so.) So I have some advantages, but I realize that unless you do something with all this information, you won't learn it. Of course, I am hoping that I can present it in such a way that you will get fired up and try out some of these ideas yourself. Like all learning, that's the secret: I have to get you to do something with it.

For years I did not want to put together a book like this because I felt that in a workshop I could provide more opportunities for active doing, that I could assure myself there that people had actually done something with some of the ideas presented. However, we found over the years that even the leaders needed more repetition and reinforcement of our messages than we could provide in one short workshop. Frankly, there is also a tendency in this field for some people to get so caught up in the doing that they lose sight of what they started out to accomplish. For them, the doing becomes the end instead of the means. Sometimes it is important to be able to go back and check up on what your doing was supposed to be doing. We hope this book will provide a basis for such an appraisal.

aCHIEVING
fOCUSED pARTICIPATION

SHARING tECHNIQUES

⊕ Organize the activity itself around sharing teams or partners

⊕ Build in formal opportunities to share discoveries or products (e.g., setting up a "gallery" or exhibit)

⊕ Use informal gatherings and set the tone by sharing things yourself (including thoughts and feelings)

⊕ Post a "Discovery Board" or map and include tear-out "Discovery Announcement" pages in the participants' Logs or Journals

⊕ Use a "Sharing Circle" to close an activity (where everyone completes the statement, "The best thing about _____ for me was . . .")

dOING tECHNIQUES

⊕ Make sure the learners know why the outcome is important

⊕ Include a participatory role for everyone

⊕ Roleplay or simulate the action involved in the understanding

⊕ Design activities that achieve both hands-on and minds-on engagement

⊕ Make sure the activity rather than the leader gets the point across

⊕ Use props to bring abstract concepts into the concrete and to serve as visual focal points (Remember: Props and Participation are Partners)

⊕ Give the learners something to do with what they have done

⊕ Add some magic and/or adventure

⊕ Reward for the behavior desired

The key to learning is doing. How many times have you heard that old phrase? No doubt, too many, but it bears repeating over and over. Only we should add a word: the key to learning is *good* doing.

Some folks appear to believe that any old doing at all meets the requirement. It's like they think as long as they put some action in there somewhere, everything will be fine. Not so. The doing needs to be focused very carefully on the intended outcome, and integrated very carefully into the whole sequence of the learning experience. You can't just tack on some active bit somewhere along the line.

The Project Wild materials could provide a case study of this particular mistake. There are numerous examples there where the action is sort of added on to a classroom discussion (action frequently represented by some "borrowed" game or task that first appeared somewhere else with different outcomes in mind).

Let's take "Ants on a Twig" as a concrete example. The objectives for the Wild activity claim that, "students will be able to 1) identify similarities and differences in basic needs of ants and humans; and 2) generalize that humans and wildlife have similar basic needs." Essentially, the activity calls for the teacher to assign small groups of students to go outside and observe ant behavior, then come together and share their results. "Included in their observations should be: evidence of how ants take care of their basic needs; description of what their basic needs are; and description of ant behavior." Obviously, the "learning" of the objectives listed must depend on what the teacher does with their reports afterward because there is no explanation for the learners about "basic needs" at the beginning.

But this is not my primary reason for selecting "Ants on a Twig" as an example of poor doing. After the "discussion" above has been completed, during which the teacher has supposedly made the points that the activity did not, there is a third step: "Now it's time to demonstrate ant behavior." Two lines of students are asked to pass each other along a narrow surface — on top of a log or stone wall, perhaps between chalk marks on a sidewalk — without falling off. (Oh yes, they are asked to use their arms as antennae as they do this.)

A CHEAP SHOT...?

I hope you won't think that we have chosen an isolated activity here just to make a point. Take another look at the supplemental collections now being peddled. I think you will find that many of their activities are either "adaptations" that really miss the mark, or paper and pencil tasks that will not really accomplish the overall objectives claimed for the materials.

Nor do we think that you are the problem. It is not the facilitators that are defective, but the materials.

EVALUATING THE DOING...

One thing you should keep foremost in mind when examining any Project Wild activity is that the objectives listed at the top of the page have no direct connection instructionally with the curriculum framework reference at the bottom of the page. Located in the Appendix, that framework, which the authors claim serves as the conceptual basis for the project, was designed as a broad philosophical overview of their beliefs. The activities in Project Wild were not designed to specifically accomplish each of those outcomes, and they have not been tested for that purpose. In fact, chances are good that the framework was devised somewhat independently of the activities in the front of the book. In other words, the framework was not put together first, then the activities developed to achieve each of its objectives. What this means is that you cannot evaluate Project Wild on the basis of those broad philosophical statements, for the actual connection between any given activity outcome and those overall objectives may be tenuous at best. You can only evaluate Project Wild activities on the basis of the objectives listed at the top of each page. As you will note, those invariably involve much simpler outcomes (and often less important ones). The reason this is worth noting here is that a lot of people read the curriculum framework in the back of the

It is pretty hard to determine what this particular step is supposed to accomplish. My guess is that going outside to observe ants (with no challenge, no context, no reward) turned out to be a bit ho-hum as an activity (don't forget though they still had everything going for them because the kids got out of the classroom to do it), and someone familiar with the adventure education activities, where this particular idea originated, "creatively" tacked it on here. In the end, someone on the Project Wild team must have realized the inherent weakness in this lack of a clear connection with the activity's objectives because they added an explanation afterwards: "Note to Teacher: Physical dramatization of concepts — in this case, ant behavior — is an excellent way to facilitate retention of concept understanding."

Of course, that really makes it worse for the discerning reader because the "behavior" demonstrated has very little to do with the objectives listed. When I used this example with some of my grad students, one of them replied that the teacher could use the activity later as a reference point for helping the kids recall what they were supposed to remember. I don't think it took her very long to realize that that was exactly the point I was making. The kids would remember the doing in the challenge rather than the point of the activity, because the doing in the actual activity was not very good doing. (And even in the challenge itself they would likely remember the doing and not some insight about ants.)

Physically dramatizing concepts is right on target, but over and over again you will see how the "Project" folks took the easy way out. They did not create activities to physically dramatize particular concepts. They borrowed activities from other sources and tried to rework them to get them to do a different job. Remember: good doing is focused doing. You start out with a clearcut idea of what you need to accomplish, then work to design an activity that gets that job done.

If you find yourself having to work very hard at getting the point across in one of your activities, or if you have to "pull it out of them," as they say, then the activity itself probably isn't very good. And chances are it is the "doing" that is off target. Go back and examine what the participants are really *doing* in the activity, and you will probably find the source of your problem.

Don't forget too that we are striving for full participation in our activities. It won't help if just a few learners actually do something while everyone else stands around and watches. (If an activity doesn't have a role for each participant, we start over.)

Finally, in earth education we like to build most of our sharing into the activities themselves, rather than tacking it on at the end. We are a bit leery of all the pop psychology that goes on in a lot of processing sessions, particularly with younger leaders. So we usually keep the sharing fairly simple and informal, and use an additional task to give the learners something to do with what they have just done. That's another nice thing about structured activities. As the learning experience moves from one step to the next, you have a better idea that your participants are with you because you can see their involvement. If you simply talk at them, you cannot see what they are doing with what you are saying.

"do EMPHASIZE tHE 3R'S: rEWARD, rEINFORCE, rELATE"

Just as the original 3R's summed up what people thought was important about education in general, these 3R's provide an easy way of remembering what we think is important for our educational leaders in specific (and that goes for the volunteers, the teenagers, and the parents too).

Of course, we should take every opportunity to build these elements into our programs, but we must also urge our leaders to make them the hallmarks of their personal contributions. There are dozens of opportunities every day where an alert leader can reward someone for a positive behavior, reinforce a point made earlier, or relate an understanding to something in the learner's experience. No amount of programming effort can do enough of this; energetic, committed, observant leaders have to provide additional hits of each of the 3R's at appropriate moments in the midst of the action.

I recall one Sunship Earth leader who would go through the kids' applications before each session looking for tidbits about each person. Then he would condense

materials and believe this is great stuff. And it's true, a lot of those statements sound pretty good. It is just too bad that the Project Wild folks didn't design a set of educational activities to accomplish them. (Hint: Anytime someone puts such statements in an appendix you can probably assume that what is up front wasn't really designed with those outcomes in mind.) I realize all this sounds like I am picking on Project Wild again, but this problem is not limited to that set of materials. It is a common characteristic of most of the supplemental collections of activities available in our field.

"Education is...hanging around until you've caught on."

— Robert Frost

his data into his own cryptic code and fit it on a single index card. Next to each student's name he would jot down a couple of items he could use as a personal entré for relating something to that kid's life — the name of a pet, a favorite sport or hobby, the number of brothers and sisters, etc.

By memorizing their names (and much of the data), then unobtrusively using his "crib sheet" whenever necessary, he could gain his learners' attention and help them personally relate what was happening to them with what they would do when they returned. (Of course, in Sunship Earth each group also returns with a whole lot of other things to work on back at school and home, but that's another level and a later story.)

Let's begin our analysis of the 3R's with how we *reward* our learners. Remember that old adage, "You get the behavior you reward for."? Well, that's what we do in earth education. We set up a situation in which we are likely to get a certain kind of behavior, then reward the learners for exhibiting it. The trick is to make sure that the activity will consistently produce the behavior you are after in the great majority of your learners.

There are all kinds of rewards (everything from a quick smile to a handful of "M and M's"), but we prefer those that will fit within the context of our experiences and work well with the other elements at the same time. (As a result, we often use overlapping rewards and reinforcers.)

For example, in Sunship Earth, at the end of the "Food Factory" station on one of our concept-building paths, the kids are rewarded with a section of an orange for serving their shift on a leaf's assembly line making sugars. Later, after applying their understanding of photosynthesis in the natural setting, they get their learning booklets stamped with the name of the concept they were just working on — energy flow. And when their counselor stamps each of these "Passports," she repeats the key concept statement, "The sun is the source of energy for all living things," thus reinforcing our overall message.

We also use multiple, interlocking rewards whenever possible. At a Sunship Study Station, in addition to receiving the immediate section of an orange after the "Food Factory" activity, followed by the Passport stamp, the students know that the Concept Paths themselves are just one part of a re-training mission that is of vital importance to the earth. Consequently, the completion of their training at the end of the week, along with the ceremonial presentation of a set of beads that illustrates their new understandings, serves as a further reward for the entire program.

rEWARDING

⊕ Give personal praise and encouragement

⊕ Provide social status and recognition

⊕ Offer tangible items or tokens (but don't let these become your motivators)

⊕ Give the learners a chance to beat the leaders

⊕ Provide opportunities for doing other desirable things later

⊕ Use the thrill of succeeding (keep the competition secondary though)

⊕ Set up levels of accomplishment (with or without awards representing the benchmarks)

⊕ Build in getting to do something uncommon or adventurous enroute

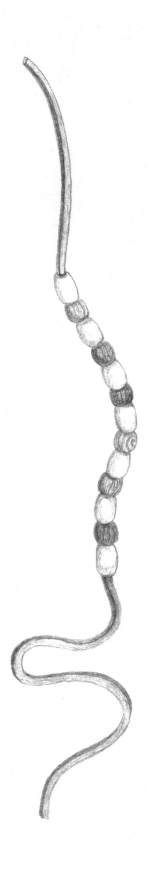

We use the term *reinforce* to distinguish the payoff for the learning (the reward) from the strengthening of it. Educators often refer to the rewards as the reinforcers, but we like to use reinforcement for specific ways of making the sought after understanding or feeling a bit stronger.

There's another old adage that might help here: "Tell 'em what you're gonna tell 'em, tell 'em, and tell 'em what you told 'em." Not that we want you to spend a lot of time telling people things, but repetition can help, especially when you have motivated learners.

In Sunship Earth we begin each morning on the Concept Path by previewing the concepts the learners will be exploring, then end the morning by summarizing the concepts they have been working on. Between activity stations on the path we ask them to further apply (and thus reinforce) each concept by finding (and pointing out with a sighting scope) another example of that understanding *in action*.

As we pointed out in the checklist on using questions, another "reinforcing" strategy we frequently use in our programs is to have visitors and volunteers play the role of reporters or inspectors (giving them special props or costumes to make them less threatening), and ask the participants to explain various activities. These "performances" are always carried out a bit skeptically (almost tongue-in-cheek), "Is that what really happens inside a leaf?", but getting the kids to explain the understandings in their own words represents first-rate reinforcement. (Please review our thoughts on questioning strategies in the section on "Things to Avoid" before trying this one though.)

In Sunship Earth we also reinforce the key conceptual understandings by using a story, "The Journey Home" (told by a counselor playing the role of an American Indian) that illustrates the same points in a different context (without even referring to them by name). Thus the participants mentally process the story in terms of its underlying points and connect those up with their previous activities.

Finally we want to *relate* the points of any program to the learners' own lives. It is one thing to understand the water cycle as an ecological system, for example, but quite another to grasp what this means for every thing you ever pour down your drain (that it's eventually going to come back out of your shower head!).

One of the problems we have worked to correct in Sunship Earth has been a lack of "bridges" between the ecological understandings and feelings gained and the students' own lives. We have disguised those systems and communities so well in our societies (the ultimate goal of the path we have taken in trying to conquer nature), that it takes a determined effort to relate each of them to daily routines.

Some of this can be done through the kind of disguise-removing that we presented in the WHATS section, but leaders need to be ready to jump in with some immediate examples right after an activity, too. "I wonder, do you think we can relate the point at the last station with what we had for breakfast this morning?"

"We have had the experience, but missed the meaning."

— T.S. Eliot

rELATING

⊕ Challenge the learners to come up with examples of things in their own lives that don't depend upon a particular natural system

⊕ Ask individuals to share stories about some of their natural neighbors at home

⊕ Ask for volunteers who can tie the point of an activity into a specific part of their daily routine (showers, clothes, meals, cars, books, etc.)

⊕ Set the learners up to take turns explaining all the energy involved in an item worn by someone, and ask the others to see how fast they can figure out what is being traced

⊕ Ask the learners to use metaphors to explain key ideas (a forest is like an apartment building, a leaf is like a food factory, etc.)

⊕ Choose something the learners like to do and have them tie it into the big four ecological concepts (energy flow, cycling, interrelationships, change)

⊕ Figure out some new ways with a group for how they can individually have a better relationship with the earth

⊕ Ask the learners to share examples of things in nature that make them feel really good

⊕ Analyze someone's daily routine in terms of what requires the most energy, water, air, etc.

⊕ Choose a plant or animal and ask the group to figure out how they are similar to it

By now you have probably figured out that a lot of the time when you focus on one of these elements you end up doing one or more of the others as well. That's okay; they lend themselves to such cross-fertilization, even if it does seem a bit much at times.

In fact, I doubt that it is possible to get too much of the 3R's worked into a program. (If the students turn green and begin looking around for a container, maybe you've overdone it a bit —but don't let the same criteria apply for the leaders, just get them some bags.)

That's another reason why real programs are so important in this field. A complete program offers almost endless opportunities for building in the kind of rewarding, reinforcing, and relating that is needed if we are going to get significant numbers of our learners to live more lightly on the earth. Programs can be permeated with these elements in ways that a couple of activities cannot. And programs lend themselves well to continual polish and refinement, finally becoming synergistic as all of the pieces begin working smoothly together.

"do model positive environmental behaviors"

Leading by example is nothing new, but nothing will probably ever come along to replace it either. Learners need to see the ideals made concrete in their leaders. As I mentioned earlier, one of the axioms of our work has always been, "if you care, they'll care." In this case, if we can demonstrate our care and concern for the earth in the way we go about living here ourselves, then we have a much better chance of getting our learners to do the same. (Participants notice if you use a blow dryer, kill spiders, carry your own cloth napkin, turn off the lights, help a worm get off the sidewalk, etc.)

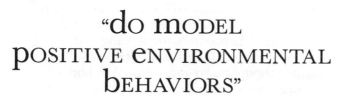

Here is a lifestyle analysis that may suggest some things you could be doing in order to live more lightly on the earth yourself. . . .

the
living more lightly
profile

Circle the response that most closely represents your lifestyle. Don't try to do "well." Just be as honest as you can. You may be surprised. Please pay close attention to your choice for each statement as the sequence of your options will vary. If an item is definitely not applicable, just leave it blank. (R = Rarely, S = Sometimes, U = Usually) Instructions for scoring appear at the end. (Please note . . . we have included some items in this list primarily to promote further thought and reflection. We hope you won't let any single statement cloud your perception about the importance of the overall profile.)

environmental habits

food consumption and packaging

I use paper towels and/or napkins	R	S	U
I take my own paper sacks (or other containers to the grocery store)	U	S	R
I avoid purchasing things in plastic containers	U	S	R
I grow some of my own food	U	S	R
I compost organic food waste	U	S	R
I use styrofoam products	R	S	U
I purchase food in bulk quantities and containers	U	S	R

I eat red meat (high on the food chain) more than twice a week	R	S	U
I avoid eating animals raised in modern factory-farm production	U	S	R
I read the labels before buying foodstuffs	U	S	R
I prepare meals without using processed foods	U	S	R
I grow or buy organically produced foodstuffs	U	S	R
I eat at fast-food restaurants	R	S	U
I belong to a food co-op in my community	U	S	R
I eat food grown locally and in season	U	S	R
I avoid snacks and other foodstuffs with lots of packaging	U	S	R
I make use of leftovers	U	S	R
I eat on airplanes	R	S	U

Other habits: _____

_____

(insert R-S-U or U-S-R as appropriate)

Impact Points   _____

hOUSEHOLD ENERGY aND SUPPLIES

I turn off electric lights and appliances when no one is in a room	U	S	R
I heat a portion of my home using renewable resources (biogas, wood, solar)	U	S	R
I run a dishwasher only when it is full and then let the dishes drip dry	U	S	R
I decide what I want from a refrigerator before opening it	U	S	R
I set a thermostat at no higher than 68 degrees during the day and 55 degrees at night	U	S	R
I use air conditioning in the summer	R	S	U
I avoid using non-essential electrical appliances (can opener, toothbrush, coffee maker, hair dryer, shaver, hedge trimmer, etc.)	U	S	R
I turn the pilot lights off on my stove	U	S	R
I check the insulation and caulking in my house and improve it if necessary	U	S	R
I keep the windows closed when cooling or heating my home mechanically	U	S	R
I make my own household cleaners out of non-toxic materials	U	S	R
I use storm doors and windows	U	S	R
I use a non-motorized push lawnmower and/or avoid mowing my lawn	U	S	R
I avoid washing clothes before they really need it	U	S	R

I wash my clothes in cold water	U	S	R
I let my washing drip dry	U	S	R
I avoid using decorative lighting	U	S	R
I use low wattage and/or energy saving light bulbs wherever I can	U	S	R
I use facial tissues	R	S	U
I use natural cleaning and grooming agents	U	S	R
I use pesticides	R	S	U
I avoid buying plastics of all kinds	U	S	R
I purchase well-made, functional clothing	U	S	R
I avoid purchasing a daily newspaper	U	S	R
I share things with my neighbors	U	S	R

Other habits: _____

_____

(insert R-S-U or U-S-R as appropriate)

Impact Points _____

WATER aND WASTE WATER

I bathe every day	R	S	U
I limit my showers to five minutes or less	U	S	R
I turn off the water heater when leaving the house for more than a day	U	S	R
I turn off the water when brushing my teeth or shaving	U	S	R

I install regulators on shower heads to reduce the water used	U	S	R
I use phosphate free detergents	U	S	R
I place something inside my toilet tank (or install a device) to reduce the amount of water used	U	S	R
I avoid pouring toxic substances or unknown chemicals down the drain	U	S	R
I purchase scented, imprinted toilet paper	R	S	U
I use naturalistic landscaping	U	S	R

Other habits: _____

_____

(insert R-S-U or U-S-R as appropriate)

Impact Points _____

tRANSPORTATION

I purchase internal combustion vehicles with more than four cylinders	R	S	U
I drive a vehicle that achieves 25 miles or more per gallon	U	S	R
I regularly walk or ride a bicycle somewhere rather than driving	U	S	R

I car pool or use mass transit	U	S	R
I keep my vehicle properly tuned and serviced for the best energy efficiency	U	S	R
I purchase radial tires and keep them properly inflated	U	S	R
I drive the same car for eight or more years	U	S	R

Other habits: _____

_____

(insert R-S-U or U-S-R as appropriate)

Impact Points _____

rECYCLING aND rEUSING

I recycle aluminum	U	S	R
I recycle paper	U	S	R
I recycle glass bottles	U	S	R
I recycle metal cans	U	S	R

I recycle motor oil	U	S	R
I use returnable bottles whenever possible	U	S	R
I reuse envelopes	U	S	R
I use both sides of a sheet of paper	U	S	R
I do not throw away items which could be repaired or reused	U	S	R
I give unnecessary clothing and furnishings to charity	U	S	R
I reuse plastic and paper bags	U	S	R
I buy throw-away pens	R	S	U
I refuse paper or plastic sacks for my purchases	U	S	R
I use disposable diapers	R	S	U
Other habits: _____			

(insert R-S-U or U-S-R as appropriate)

Impact Points   _____

NATURAL CONTACT AND RESPECT

I visit or take a walk in a natural area each week	U	S	R
I notice the changing phases of the moon	U	S	R
I share my love of nature with others	U	S	R
I seek support within my spiritual views for living more lightly on the earth	U	S	R
I pay attention to the natural changes in the seasons	U	S	R

I make an extended visit to a natural setting at least once each year	U	S	R
I notice the color of the sky	U	S	R
I treat all living things with respect	U	S	R
I eat baby animals (veal, lamb, etc.)	R	S	U
I bell my cat	U	S	R
I kill things for recreation	R	S	U
I practice minimum-impact techniques when I go camping	U	S	R
I purchase products made from wild animals	R	S	U
Other habits: _____			

(insert R-S-U or U-S-R as appropriate)

Impact Points   _____

MISCELLANEOUS

I engage in (low energy) recreational activities   U S R
I spay or neuter my dog or cat   U S R
I work at learning more about ecological processes
    and what they mean for me   U S R
I examine any financial investments I make in terms
    of the environmental impact produced   U S R

I purchase simple, durable, low energy things
    whenever possible   U S R
I work consistently at improving upon my own
    habits   U S R
Other habits: _____

_____

(insert R-S-U or U-S-R as appropriate)

                         Impact Points   _____

ENVIRONMENTAL PARTICIPATION

I discuss pending environmental legislation with
    people around me   U S R
I ask my workplace to engage in more
    environmentally-sound practices   U S R
I help restore natural areas   U S R
I keep abreast of current environmental issues   U S R
I actively support an environmental action group   U S R

I inform my elected officials about my environmental
    concerns and recommend actions   U S R

I contribute 1% or more of my annual income to
    environmental causes   U S R

Other actions: _____

_____

(insert R-S-U or U-S-R as appropriate)

Impact Points   _____

STANDARD ADDITIONS

Number of rooms in my dwelling	1-3	4-6	7 or more

Number of automobiles I own	0	1	2 or more
Number of recreational vehicles I own	0	1	2 or more
Number of houses I own or rent	0	1	2 or more
Number of miles I travel to work	0-4	5-14	15 or more
Number of medium to large pets I own	0	1	2 or more
Number of pounds I am overweight	0	10-20	21 or more

                              —   ——   ————

Impact Additions: _____

REPRODUCTION SURCHARGE

Number of children my spouse and I have produced
    multiplied by 200   _____

STANDARD SUBTRACTIONS

I wash my dishes by hand (5 points)   _____

I use a composting toilet (5 points)   _____

I buy used clothing or make my own (5 points)   _____

I wash my clothes by hand and let them drip dry
    (5 points)   _____

I do not use an air conditioner to cool my home
    (5 points)   _____

I have made arrangements for a natural burial
    (5 points)   _____

I do not own a car (10 points)   _____

I buy very few material things (10 points)   _____

I teach my children how to live more lightly
    (10 points per child)   _____

I have relatives, friends, etc. living with me
    (10 points per person)   _____

Total Subtractions: _____

SCORING

First, go back and tally up the number of points you have
for each section. Circles in the left column are worth 0 points,
those in the middle column are worth 2 points, and those in
the right column receive 5 points. Enter the total points for
each section below.

Environmental Habits

Food Consumption and Packaging   _____

Household Energy and Supplies   _____

Water and Waste Water   _____

Transportation   _____

Recycling and Reuse   _____

Natural Contact and Respect   _____

Miscellaneous   _____

Environmental Participation   _____

Total Impact Points: _____

Second, tally up the number of points you should add to the
above totals. In the Standard Additions section circles in the left
column are worth 0 points, those in the middle add 10 points each,
and circles on the right gain 25 points each. Next, figure out the
surcharge for the number of children you have produced. (We
don't mean to denigrate the value of human life, but that's probably

the greatest impact you will personally ever have on the planet). To do this you should multiply the total number of children you have produced by 200 then divide by 2 (for your spouse's share). Finally, total up the number of points you have under standard subtractions. Now you can enter the totals below and finalize your score.

Environmental Habits and Participation   _____
Standard Additions   _____
Reproduction Surcharge
     (100 points per child)   _____
Standard Subtractions   _____

Your "Living Lightly" Score   _____

We know this final score doesn't mean much (you would probably be a fanatic living alone somewhere in a cave if you ended up with a perfect score for living lightly, i.e., with no points at all), but we thought the profile might give you some ideas for how you could improve upon your own lifestyle.

Naturally, all the numbers we use are arbitrary, but they may give you a basis for comparing yourself with others. Ask your friends or colleagues to score themselves, then talk about your respective profiles. Of course, in this case you may want to figure your score differently for different comparisons. For example, if you would like to compare your daily impact with someone who has produced less offspring, then just match their points in that category and compute your score again.

The main idea is to begin thinking about ways of lessening your impact on the earth (relative to where you are now, not where someone else is). The truth is most of us in the western world have such an incredible impact upon the planet that it is almost obscene to include even some of our commonplace habits on a list like this since our conspicuous consumption, in comparison with the lifestyle of most people on the earth, makes us appear almost callous and shameful. Nonetheless, we have to begin where we are. The important thing to remember is it is easy to lower your impact on the earth, and we can all work to "live more lightly."

Duplication Masters for the Living More Lightly Profile can be purchased from The Institute for Earth Education.

Whew! This section took more words than I thought it would. Please remember, just as the map is not the territory, these guidelines are not the program.

They are some means, not the ends. And we know you will find situations out there that our guidelines don't cover. Follow your heart. If you have read this far about our work, it is probably a good judge.

In the next section we are going to look at one of the major differences between earth education and environmental education, i.e., how we get our participants out there and "immerse" them in the natural systems and communities of the earth.

iMMERSING

"We believe in including lots of rich, firsthand contact with the natural world."

In <u>The Earth Speaks</u> I shared a story about St. Francis of Assisi who suddenly began ringing the church bell late one night in his village. When the awakened townspeople rushed to the tower to see what was wrong, St. Francis called down to them to look up, right then, at the beautiful moon overhead. Shut up inside their homes they had been missing one of the earth's supreme treasures.

We need bellringers like this in earth education. We need to awaken people to the marvels and mysteries of daily life that they are missing. However, there is a bit more to it than that. If we are going to help people build a joyous relationship with the earth again, we have to accomplish three major objectives. First of all, like St. Francis, *we have to get people out there*, only a bit farther away. You cannot accomplish what we are talking about inside most cities. You have to be "immersed" among the wild and growing things, in direct contact with the elements of life. Next, *we have to help people take in more of what's around them* out there. Lots of folks have gone someplace where they were closer to the natural world, but never really touched, or were touched, in turn, by it. Walking along on an asphalt path chattering away about something else will not get the job done. And finally, *we have to make sure people have a good time* while they are there. We want people to remember their experience as a joyous one. It will not do us much good if they decide it was not fun, and they don't want to go back.

In <u>Nature and Madness</u>, Paul Shepard proposes that human beings are genetically programmed for a period of immersion in the natural world during infancy, and if they are deprived of that experience, they will never develop a good sense of relationship with the earth. (That is what we are going to do in our "Earthborn" program.)

"PERSONAL IMMERSION"
(GETTING PEOPLE OUT THERE)

A number of years ago, I received a phone call from a television reporter in Boston. He said he had heard about our work and wanted to tell me about something rather startling that had happened to him. One day, while conducting some "on-the-street" interviews, he decided to ask people about the sky above them. So he began stopping folks, holding his hand above their eyes so they could not look up, and asking them to describe the sky overhead. He said he was stunned to find that most of them couldn't do it. Time and again he would stop people walking along in the open air and find that they couldn't describe the natural canopy under which they walked. He was obviously shocked by this phenomena. I was not. In fact, I was a bit surprised that he *was*. Most large cities are like gigantic colonies of animals, and many of us have lived so long now among those artificial canyons, dominated by vast indoor marketplaces, that we have literally lost sight of the reality of life on earth.

Of course, in many of our cities today the light overhead is so often little more than a uniform glare emanating from nowhere in particular as it filters down through the layers of smog hanging above, that it is no wonder people no longer notice the sky in such places. Did you know that many of the children in our largest cities have never even seen a rainbow?

Sometimes our urban settings remind me of the analogy of Plato's Cave, which proposes what life would be like for people who had always seen shadows. Picture a large underground cavern with a thin section of rock like a raised roadway running through its middle. A tunnel at either end of this "road" connects it with the surface, and the people living nearby use this underground passageway as a shortcut between their villages. Imagine that on one side of that central rock ledge a fire was always kept burning so that its light would cast the people's shadows onto the opposite wall as they passed through the cave. Since the villagers would be carrying various objects with them, many of the things in their daily lives would be illustrated by the shadows cast on that rocky surface. Now, pretend that below this central rock bridge some people have been chained up in such a way that during their entire lives they had never been able to see anything except the shadows appearing on the wall of the cave before them. All they had ever known about other life was contained in those flickering, moving, talking shadows on that wall.

What do you suppose would happen to these people if you suddenly went into the cave, struck off their chains, and hauled them out onto the surface? Forget the blindness caused by the light, etc., what would these people perceive as real? Wouldn't actual objects probably be seen by them as illusive and their shadows as genuine?

Well, I suspect that is what has happened in our cities. We have lived with the artificial and synthetic for so long that we have begun to believe they represent the reality of life here, while the earth's natural communities and systems have become mere shadows for us. Given this ingrained perception, it is going to take a fairly dramatic experience for many of our learners to begin changing their attitudes about the natural world. (And in terms of impact the result will probably be similar to what those cave people would have experienced if they were suddenly plucked from their caverns and exposed for the first time to a world beyond their boundaries.)

To immerse means to plunge in or to absorb deeply. In earth education, we have to achieve both parts, i.e., we have to get people to plunge in and absorb deeply. When it comes to natural experiences many of the people in our cities are like those highly compacted sponges they sell in the novelty shops. You remember, the kind you add water to and then watch them swell up into a regular size sponge right before your eyes. Well, lots of city dwellers are like that, they have been out of touch with the earth for so long that you have to plunge them in first in order to bring them up to an appropriate size. Only then can you begin working with them on absorbing ever more deeply the various facets and underpinnings of the natural world that supports them.

Let's look at another example. Most leaders would accept that you cannot take someone from another culture and immerse him in yours on one occasion and expect him to go away with the same attitudes towards your cultural systems that you have developed

The Ecology of Imagination in Childhood is the masterwork of Edith Cobb, who spent much of her life exploring the idea that the cosmic relationship with the earth lies at the root of human development. In examining the lives of hundreds of creative thinkers from around the world she discovered that during early childhood each of them had undergone an acute, sensory response to the natural world — a sudden awareness that being and becoming are part of the flow of life, and a sense of their oneness with it — and each reported that the experience had been a life-changing one for them.

over a lifetime. Well, the same holds true for natural systems. Those attitudes must also be developed through repeated immersions.

The point is, taking a group of youngsters on a field trip each year, then returning them to the rather sterile confines of most schools, will simply not work for what we must achieve. Anyone who has ever traveled to a foreign land knows that you cannot begin to grasp the cultural patterns of another people through one quick visit. You need repeated exposure to many different aspects and functions of a culture, plus a chance to reflect upon what has happened to you.

Consequently, earth educators must get their learners out there in touch with life over and over again, and at the same time help them make sense out of what they are doing. As Aldous Huxley put it, experience is not what happens to you, but what you do with what happens to you. So when we get our learners out there we also want to give them something they can follow through with later when they return to their classrooms and homes. And that leads us to our second objective.

"ENRICHED PERCEPTION" (HELPING PEOPLE TAKE IN MORE)

Why is it that we insist on perpetuating the myth in our societies that being childlike and childish are the same thing? They are not. They are totally different things. What could be better than holding on all one's life to some of those marvelous characteristics of children? How about their bubbling, effervescent joy at being alive; their marvelous ability to totally immerse themselves in the moment, to lose track of time and space and just merge with the flow of life; their insatiable curiosity and sense of wonder about everything around them? Ah, to be able to discover your toe again at fifty and behold it as if for the first time.

We get so puffed up with our own self-importance that we lose the ability to get down on our hands and knees, to see, close-up, some of the wonders of life. We are unwilling to peel off our fancy costumes and grab onto life by the handful. As affluent members of western societies, we have lived too long swathed in energy, ensconced snugly in our synthetic cocoons. And today, this reluctance to partake appears ever earlier in our lives.

Someone shared a story with me at a workshop that illustrates the problem rather well, I think. A somewhat bored teenager staying with her grandmother in the country went each day down the gravel lane to check the mailbox. Since she always returned still focused on the mail she had or had not received that day, one afternoon her grandmother gave her a pail to take along with her with the request that she bring back something beautiful. But the girl replied, "Oh, Granny. There's nothing out there but weeds and stuff. It's just an old path." Nonetheless, her grandmother persevered saying she was sure the girl would find something of beauty to bring her. Of course, she did. The first day it was just a few berries, the next, some colorful leaves, but then came rich-smelling toadstools, deeply veined rocks, curly pieces of vine, and assorted other natural wonders. The grandmother's simple directions, an important symbolic tool, and her obvious delight at sharing the discoveries, gave a young girl a whole new dimension to her world, and a new appreciation of natural beauty.

"A day spent without the sight or sound of beauty, the contemplation of mystery, or the search for truth and perfection, is a poverty-stricken day; and a succession of such days is fatal to human life."

— Lewis Mumford

Like that over-stimulated teenager, many people visit the natural world today, but return empty-handed. Their perceptions are so clouded, or they are so full of themselves, that they miss much of what surrounds them. As if emerging from a cocoon, they come to the natural world wrapped in layers of gauze. In earth education, we have to help them peel away those layers, to reach out to the world with a fresh perspective, and like a young child to look long and lovingly upon the earth, to explore again the smallest interstices of life. And not just with their eyes either. Helen Keller could neither see, nor hear, but oh, how she could feel:

> What a joy it is to feel the soft, springy earth under my feet once more, to follow the grassy roads that lead to ferny brooks where I can bathe my fingers in a cataract of rippling notes. . . .

To rediscover the earth and renew our relationship with it we need to rebuild and refine some of the natural skills and senses that we have let atrophy inside our artificial cocoons. In <u>Acclimatizing</u> we included eighteen exercises for building such natural awareness.

These simple activities were broken down into four broad areas of skill: sharpening senses, seeking patterns, perceiving wholes, and distilling essence. We have capsulized them in the checklist below to give you an idea of the breadth of the task and the variety of tools available.

ɴATURAL ᴀWARENESS SKILLS

SHARPENING SENSES

☑
- ☐ Touching — touching with your hand, but feeling with your body
- ☐ Tasting — eating with all of your senses
- ☐ Hearing — listening to the purity of sounds
- ☐ Smelling — tracing the smells around you
- ☐ Seeing — channeling other senses through your eyes
- ☐ Feeling — paying attention to your feelings

SEEKING ᴘATTERNS

- ☐ Focusing — concentrating on gathering many clear images
- ☐ Framing — enclosing scenes which you want to examine
- ☐ Grouping — looking for arrangements in forms and lines

ᴘERCEIVING WHOLES

- ☐ Expanding — observing the overall aspects of a scene
- ☐ Filling — accentuating the negative
- ☐ Surveying — examining things from varying viewpoints
- ☐ Observing — letting the natural world engulf you

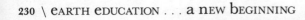

☐ Orchestrating — using all of your senses to fuse the facets of awareness

dISTILLING eSSENCE

☐ Scrutinizing — looking for the small things

☐ Empathizing — roleplaying natural qualities

☐ Silencing — working at turning off that voice in the back of your head

☐ Waiting — becoming an empty vessel waiting to be filled

All of these exercises can be used to help individuals cleanse their perceptions as they immerse themselves in the natural world. Nonetheless, they are still rudimentary exercises, the place to begin. However, for many years in our Advanced Acclimatization Workshops, we worked on another set of "senses" which combined to form the affective side of that special sense of relationship with the earth that we always had as our primary goal. Here are four additional "senses" we believe we must also engage and sharpen:

SPECIAL SENSES

SENSE OF WONDER

☞

☐ looking upon the world with habitual awe

☐ maintaining curiosity about life's comings and goings

☐ reveling in the unusual and unexpected facets of the commonplace

☐ exploring and seeking in unselfconscious ways

SENSE OF PLACE

☐ being at home in a natural community

☐ feeling caught up in the synergy of an area

☐ seeking new dimensions and forces in a familiar place

☐ absorbing over time the textures and moods and qualities of a piece of the earth

SENSE OF TIME

☐ comprehending the continual passage of time as a constant of life

☐ seeing being as becoming

☐ feeling connected to all past and future things through the flow of life

☐ grasping that each piece of the earth and particle of life has a story in time

SENSE OF BEAUTY

☐ enjoying the color and form, pattern and texture of natural things and scenes

☐ perceiving the harmony in healthy natural communities

☐ noting the continual changes in light and shadow

☐ appreciating the rich diversity of life in the natural world

☐ seeing "the world in a grain of sand and heaven in a flower"

If perception is the ability to mentally grasp things through the senses, then the special senses above are like perceptual peepholes on our world. They enable us to glimpse a whole new dimension of our place in space. And just as we commonly use exercises in life to aid us in developing a skill, we can do the same here. As leaders, we can sharpen all of our special senses through regular practice, and thus be prepared to model the ecological feeling that is at the heart of earth education.

Most of all though, that practice needs to include more nature "up against the skin." You just can't achieve the kind of perception we are talking about all wrapped and bundled in the latest synthetic material like so many space explorers. You have to get out there and squish it between your toes, get it under your fingernails, and feel it in your hair. Don't let your relationship with the earth end up in the category of a fair weather friend either. Get out there and suffer a bit with it too.

One final note: In earth education we have devised lots of activities for helping people get more in touch with the natural world. (After all, we're the original "tree huggers.") But the real secret of our success often lies in how we present those activities at the beginning of a program and then how we "re-present" them ourselves during the course of a regular day. One of our Sunship Study Station leaders likes to tell the story about how a rowdy group of boys were dramatically influenced one day by their high school counselor, who they all looked up to because of his high energy, adventuresome ways, and friendly spirit. As the kids tumbled out of their cabin on the second morning of the session, charging off with their counselor on some daring mission, he suddenly stopped, calling out, "Hey, wait a minute, Hold on! I just have to get a hit of this flower." His personal example (albeit carried out in a somewhat exaggerated way) soon had all the boys in his group stopping to smell the flowers.

iMMERSING tECHNIQUES

⊕ Change your vantage point to make the familiar unfamiliar

⊕ Use all your senses in exploring an area

⊕ Take off some clothing to increase physical contact

⊕ Simulate natural processes (flow like the water, leap like the wind)

⊕ Crawl or roll or float instead of walk

⊕ Take away your sense of sight or sound to heighten your other senses

⊕ Roleplay other creatures and things

⊕ Look at the world from the perspective of something else

⊕ Get off the path, go cross-country

⊕ Find dramatic places to perch

⊕ Pet and hug and kiss things

⊕ Go out in unusual hours and conditions

⊕ Play with a childhood toy, then go for a walk in a natural area

⊕ Spend a whole day outside (dawn to dusk) and imagine that it is the last day of your life

Please don't confuse these assorted techniques, nor the previous natural awareness exercises, with a program. They can support, enhance, or extend, but they are not the main event. A bunch of these activities will remain just that — a bunch of activities. We have included them here primarily to give you some ideas for what you can do personally to increase your own contact and improve upon your own perceptions, but we suspect you will also find them useful at many different points during a complete earth education program.

"PLEASURABLE EXPERIENCES"
(MAKING SURE PEOPLE HAVE a GOOD TIME)

Regrettably, we have fostered another terrible myth in our societies, that is, that names and numbers are real, but feelings are not. I suppose it is because we can measure the former, while the latter often elude our attempts at quantification. However, it doesn't take any great perceptive ability to figure out that people pursue things in life because they *feel* good about them. Be it people or places, possessions or professions, we seek what we feel good about. In fact, for most people, the last time they will ever pursue anything for very long that they *don't* feel good about will probably be their last days of school.

Feelings are certainly real and just because we find them difficult to assess should not lead us to overlook their importance in people's lives. The leader who ignores the feelings of his learners does so at his own peril. (Do you suppose that may have something to do with why over 25% of American teenagers drop out of school?)

Sadly, even though these points seem self-evident, I have to report that we have let some people loose in our schools who still believe that learning should be pain. For these folks, the brain seems to be a muscle to be exercised and the classroom serves as the equivalent of some contemporary health club exercising machine. "No pain, no gain" has become their motto for learning.

Please don't misunderstand what I am saying. We don't believe all learning should be a lark either. In fact, chances are that some of the most strenuous learning people undertake will involve something they feel good about doing. That is why they are willing to work so hard at it. So having a good time is just one part of our criteria, though tragically, lots of outdoor programs seem to conclude that it is the whole point. It's not. But overall, we believe learning should be viewed as an adventure, because life itself is an adventure. And when it is all said and done, daily living is learning, and lasting learning is living.

This is why the stress-challenge philosophy so popular in outdoor education, even when administered well, may have some inherent drawbacks for building a relationship with the earth epitomized by joy and harmony. We don't want our learners to come away feeling like they have conquered either themselves or the natural world. We want them to come away feeling so good about the rich, firsthand contact they have had with natural communities and processes that they fairly tingle with excitement at the prospect of getting back out there.

I don't want you to think we have never made this mistake ourselves either. (Lots of things in this book were learned the hard way.) Years ago, at the northwoods camp where our work began, we had something we called a "swamp tromp" to introduce the campers to an unusual area of camp they had never seen. Somewhere along the way we got to billing it as a sort of macho, can-do experience. The kids would step off fully clothed into the lake right in front of their cabin, then wade in and out along the shore for about a mile. It offered a perspective on both the lake and forest and intervening marsh communities that few of them had ever experienced. Anyway, we usually scheduled this by cabin groups, but one day the counselor for a senior group came to tell me his kids were holed up in their cabin and refused to go. I was in a hurry for something else, who knows what, and zipped down to their cabin to roust them out.

When I stormed in and proceeded to read the riot act to them, one of the campers ran out and the counselor chased off after him. Meanwhile, I pulled one kid off of his bunk and yelled at the others that they sure as hell were going to go on this experience. (No, that's not what we mean by pulling instead of pushing your learners.) Actually, I shouldn't make light of this. In retrospect the whole scene is one of those excruciatingly embarrassing moments that are almost painful to recall. Even after all these years, I can still barely tell the story. I just don't know what came over me that day. It wasn't until I saw the counselor returning with a crying, kicking youngster over his shoulder that I suddenly came to and realized what I was doing. It was a devastating experience for me. I had forgotten rule one: make it fun.

I doubt that my apologies got me anywhere with those boys — the damage was done — but I never forgot the lesson I had learned myself. You can't force your learners into feeling good about the natural world. You have to start out with the idea that you are going to have a good time, and make sure everything you do feels like it is.

hAVING a gOOD tIME

Here are a few hints for how to make sure your participants have a positive experience. Most of all keep in mind that your own enthusiasm will be contagious.

⊕ Expectations — focus on the appealing parts of the experience (remember to pull, not push). Build up to the event over a period of time (send them notes, maps, photos, etc.). Share some special experiences you have had yourself, but don't promise more than you can deliver.

⊕ Clothes — ask everyone to bring a set of "animal clothes" to wear whenever you go "immersing." Keep them in a special bag or hang them on a special hook.

⊕ Tools — everyone likes to have something to use, something to take along when you go exploring (bandannas, binoculars, discovery packs, journals, etc.). Just make sure the tools don't get in the way of the objective.

⊕ Fears — accept them, unconditionally. Don't try to talk people out of them. Ask your participants individually if there are any "worries" that you should know about so you can help out. Explain that no one should be embarrassed because everyone has some concern like that. Thank the people who tell you something, and ask them if there is anything you can do to make it easier for them to deal with the problem.

Before we finish this section, let me recap the main points of our second WAY. In earth education we believe in immersing people in the natural world . . . over and over again. And to be successful we have found there are three things we have to focus on: we have to get them out there; we have to help them take in more while they are there; and we have to do everything possible to make their experience a joyous one.

In the next section we are going to examine how the individual learner relates both *with* and *to* these experiences and their settings.

"It takes a very long time to become young."

— Pablo Picasso

rELATING

"We believe in providing individuals with time to be alone in natural settings where they can reflect upon all life."

Another way we could have stated the third one of our WAYS would have been to say we want to provide individuals with time to be alone in natural settings where they can reflect upon all life, *including their own.* In other words, we want to help our learners relate both *with* and *to* the other life of the earth. We want them to relate *with* it personally on an affective level, relate *to* it individually on a cognitive level, then examine their own lives in light of both experiences. (As we saw in the chapter on Acclimatization, it's feeling the processes and processing the feelings.)

We use solitude experiences to help us accomplish this interplay between the understanding, the feeling, and the processing. And we often suggest that our participants may want to spend a portion of such time writing in a special journal or log or diary to help them sort out some of their thoughts and feelings about their own relationship with the earth. However, even though solitude experiences are an important "way" of processing things, we also want to make sure our learners don't spend all of their time out there locked in on some head trip either. I know that must sound confusing, especially in terms of what I just said about writing in a journal, but keep in mind the dual purpose of these experiences. To accomplish our goal of helping people relate *with* the natural world we are going to have to help them get out of their heads (and into their surroundings) during some of their time out there. Let me explain what I mean. Stop right now and take just a moment to really listen to the sounds around you. C'mon, really do it.

Chances are good that there is this little voice in the back of your head trying to name the things making those sounds. I call it the little reprobate that lives in the attic of your mind. Oh, the psychologists have a fancy word for it, but the little reprobate will do for our purposes. I want to suggest that that little voice has the power to both illuminate life for you and to insulate it, and it can do both equally well.

"Feeling the processes and processing the feelings."

What did it just do in this case? I said, stop and listen, really listen. But the little voice jumped in between you and the sounds and started trying to name the things making them. In a way it got in between you and the world. It insulated you from the actual sounds around you. If you want to get a handle on the little voice in situations like this, then force the little voice to take letters of the alphabet and put them together to duplicate the sounds you hear. That way you will have to really tune in to the sounds themselves. Why don't you try that technique out right now? Just stop reading, tune into a sound nearby, then force the little voice in your head to use letters of the alphabet to duplicate that sound. Go ahead, try it. I guarantee you will never hear things quite the same way again.

Another example of the distracting nature of the little voice that I imagine most people have had is the experience of driving down a highway somewhere when you suddenly seem to come to at the wheel. Do you know what I mean? It's like you have been driving for a half an hour and you don't even remember where you've been. You usually hold on to the wheel a bit tighter in such instances, like that's going to help at that point. Again, in a sense, the little voice had you somewhere else. It had gotten in between you and life.

In one of my speeches, I tell an embarrassing tale about myself. I'm a junkie . . . an apple dumpling junkie. I don't know about you, but I crave apple dumplings. I'm addicted to them. I remember one day years ago when I was so excited because we were going to have apple dumplings for dessert, and I had been looking forward to it all day. Anyway, when dessert arrived that evening, I was all fired up in anticipation, but then, the next thing I knew, it was gone. It was like my apple dumpling had gone from my bowl to my belly and I didn't even remember the route. All I could do was to stop, hold my stomach, and go, "Oh, no!" Sadly, that little voice in the attic of my mind had taken me off somewhere else, and I didn't even remember eating my favorite dessert. That is when I first began to focus on the problem of the little reprobate, I think. It had gotten in between me and my apple dumpling, and that was a serious problem.

Unfortunately, many people go through much of their lives in that fashion. Their little voice has them locked into thinking about the past or the future much

"'Now, I give you fair Warning,' shouted the Queen, stamping on the ground as she spoke; 'Either you or your head must be off, and that in about half no time!'"

— Lewis Carroll
Alice in Wonderland

of the time. Consider your own life. How much of your day is spent thinking about the past, or thinking about the future? In fact, what is probably the most useless conjugation in the English language? I would nominate, "should have done." It's gone. It will never return, but we spend a lot of time thinking about it. Or how much of your day is spent thinking about what's coming: "Hey, I'm going to be cool tonight." But that experience may never come either (and often doesn't), even though we spend a lot of time thinking about it, too.

You see, I think a lot of people play out their lives like they were the ball in some sort of cosmic ping-pong game. They bounce around over on one side of the table thinking about the past for a while, cross the net for a moment, then bounce around on the other side of the table thinking about the future. Only once in a while do they reach out and actually grab on to the world right now, the net of the present.

Please don't misunderstand. I am an historian by training. I think it is important to think about the past. And I am a futurist as well. It's vital; we had better think about the future. But let's do it. Let's sit down and think about the past. Let's sit down and think about the future. Let's just not get so caught up in the ping-pong approach to life that we forget who and where we are much of the time.

Ask someone sometime to point to "me." That's a word we use all the time, so ask someone to point to the me they mean. Do you know where a lot of people will point? To that little voice that lives in their head. And if you ask about the rest of them, they are likely to look down and respond, "Oh, that's my body." I love it. How can you say that you have a body? You don't *have* a body. You *are* your body. From the tip of your toes to the top of your head . . . that's you, the whole you. However, a lot of people walk around like the rest of them is just a vehicle for their head to ride on. You watch. You'll see them. They have literally lost touch with themselves. They are trapped in that ping-pong game of life, and it is going to take a lot of work to get them back in touch with the earth again.

Folks, do you know that you do more every single day that you don't think about doing than you do that you do think about doing? That's true . . . think about

One of my favorite stories about this problem was told by a Danish mystic. He explained that God sent his angel, Gabriel, down to earth to find someone who had just a moment of time. In return, God said he would give the person immortality, but Gabriel returned the next day empty-handed. "I don't understand," God said. "All I need is just a moment of a man's time, and in return, I will give him eternal life." But Gabriel explained that no one down on earth had a moment of time. They all had one leg over in the past and one leg over in the future, so nobody had a moment of time right then.

it. Right now, you are doing a thousand, thousand things and you are not thinking about any of them. It's a good thing too. What if after breakfast every morning we all had to go sit in a corner somewhere and say, "Okay, liver, go ahead, do it?" We are the most incredible organism ever spewed up in the film of life, and it all just works.

Let me share a story with you about a young man who got out of his head for a while, and returned much richer as a result. For many years in our original Acclimatization program, I would gather up some of the ACC staff, plus a couple of older campers, and after our camp season ended in August, head for Isle Royale National Park up in Lake Superior.

During the trip the ACC folks would end up rapping a lot about what had happened during the summer and coming up with new ideas for the next year. I noticed that whenever we got wound up in these conversations, Chris, one of the older campers, would always hang around taking it all in. (Actually, Chris had not been at camp for a couple of summers. He had been off seeing the world instead, and had called up at the last minute to see if he could join up with us in going to Isle Royale.)

Although Chris had already seen the world at fourteen, I don't think he had ever felt it until one evening when we were camped on a small island offshore from the mainland. It was one of those wet, grey days when heavy, leaden clouds hang just overhead, spitting rain at you from time to time. The gulls were all huddled on the sandbar out front, while we were holed up in our shelter. The talk had come around that afternoon to one of our solitude-enhancing experiences which we call Magic Spots. Chris seemed to be absorbing all of this as usual so I asked him if he would be interested in trying out some of these ideas. He nodded enthusiastically, so we explained what a Magic Spot was all about, how it sort of finds you, then sent Chris off exploring.

It wasn't long before he returned, saying that he had found a great spot along the shoreline, but he wanted to get a journal like all of us had been using. Someone rummaged around and came up with a spare journal, and Chris headed back to his spot. Within a half hour or so the sun began setting, and

WIGGLE YOUR TOES...

Right now. Go ahead, do it. This is a participatory section. Lots of people even tip their heads down when they do it to check out what's happening down there. You ought to do that about once every hour just to keep both ends in focus. Whatever you do, don't get caught in the ping-pong approach to living; reach out, grab onto the earth and feel your oneness with the flow of life.

"All the miseries of mankind come from one thing, not knowing how to remain alone."

— Pascal

the rays shot out suddenly from under all those clouds and just bathed everything in a golden light. The gulls took off again and as they turned and banked overhead the sunlight would catch them so they flickered off and on like large flakes of gold in the sky.

When Chris finally came back, it had grown dark, and we were all sitting around a small fire, talking about the day. He just stood there like a young pup, holding his journal, and obviously pleased with himself. We never force kids to share their journals, but sometimes we ask them if we could read or hear some of their thoughts. So I asked Chris if he had something he would like to share with us. You could tell that he was excited, and he proceeded to read us a description of the closing time of that particular day — a description that really captured the sounds and sights and smells, the feel of those moments in the flow of life.

In a way that is the end of the story, until November came and Chris called one evening to ask if I was going to make my usual winter pilgrimage to the everglades. When I replied in the affirmative, he asked if he could go along. I said sure, and we got to talking about Isle Royale and how different the everglades would be. All at once, Chris said, "Hey, do you know this Thoreau guy?" When I responded that yes, I thought I knew who he meant, Chris said, "Well, my English teach was talking about him, and I told her we were all doing that stuff up at Isle Royale last summer. Anyway, she didn't believe me, so I took my journal in and it just blew her mind, and I'm gettin' A's in English."

I think that Magic Spot at Isle Royale represented not only the first time Chris had actually felt the flow of life, it also turned out to be the first time he had connected up his English class with the joy of writing.

As Chris's experience illustrates, if we are going to succeed in earth education, one thing we must help people do is to get out of their heads and into nature. Reflection and analysis are important but we need to make sure people have something to reflect upon first. I guess you could say that solitude time is for both getting into your head and getting out of it, for both reflection and engagement.

In summary, for the kind of "Relating" we are interested in here, there appears to be a regular, though unusual, progression of events. First of all, we have to help people get away from all the man-made static. It is our belief that a "natural" synthesis of individuals and communities partially emerges from intense immersion in a wild setting. Next, we have to help them get out of their heads, and here's that ironic part again, so they can get back in touch with themselves. We don't believe a person can achieve the clarity of perception we are after unless the little voice is stilled for a while. Finally though, we have to guide the little voice into new pathways of reflection. Since experience is not what happens to you, but what you do with what happens to you, we have to nudge people into assimilating their feelings for the earth and their new way of seeing it, by incorporating those into new patterns in their thoughts and actions. And that takes some forethought and planning.

Here are some examples of the kinds of tasks or questions we might include in a special booklet for the learners to use during the daily solitude time that we include in all our programs.

Ernest Thompson Seton was a North American nature writer and artist at the turn of the century who based his stories and paintings on the countless hours he spent patiently observing the natural world. In over 40 books he brought to life many of the common wild creatures of the continent and inspired a whole generation of nature lovers in the process.

INPUTS FOR SOLITUDE JOURNALS

☐ "Do some 'Seton-Watching' to feel the flow of life around you. First, get really comfortable, maybe sitting with your back against a tree and your legs stretched out. Fold your hands and put them in your lap. Then take a couple of deep breaths and let them out slowly. Begin to relax, and try not to move at all. Just wait like this, really quiet and still, until the things around you start returning to their normal activity."

☐ "How Am I Doing?" Jot down three things you have learned to do in order to have a better relationship with the earth:

1.

2.

3.

☐ "Listen to the sounds around you. Try not to name them, just listen. Think of them as instruments in an orchestra. Hear the patterns, the solos and the individual instruments playing louder and softer. . . ."

□ "Complete this statement: Today, at my Magic Spot. . . ."

□ "Imagine what it would be like to be one of the plants or animals around you. Really try to look at things from its point of view and discover what life is like for it. Try writing an entry for one day in its 'diary.' 'Today, when the sun came up . . .'".

See our chapter on "Magic Spots" in <u>Sunship Earth</u>, our "Log Book," and the "Earthkeepers Diary" for additional ideas.

A SPIRITUAL DIMENSION...

There appear to be many doors out there in the metaphorical wall of cosmic consciousness, and in earth education we want those coming through any of them to feel welcome. We believe that the spiritual dimension of building a better sense of relationship with the earth is an important one, and we would hope that environmental leaders everywhere could find support for earth education within their own religious beliefs and traditions. Please help us encourage them to do so.

In the end, one of the best ways to set your learners up for their solitude experiences is to share with them a bit about your own. Some personal stories (perhaps even showing them a favorite Magic Spot), will help them get the idea of what these special times are all about. Consider letting them take a look at your journal, or maybe read a short entry to them. Modeling the behavior yourself, and sharing your experiences, will do a lot to motivate your learners.

An elderly lady in an autumn workshop I conducted years ago explained after a solitude experience that she had been a bit skeptical about all this getting out of your head stuff and merging with the flow of life. However, she decided to go along with it and found a place to sit under a tree in the forest. After she had settled in for a few minutes, she said she was suddenly aware that it was actually raining leaves. "They just kept falling. Hundreds and hundreds of them. They just kept coming. After a while, I wasn't sure if I was me or one of those leaves." As an earth educator, that's one of those rewards you will receive from time to time for your persistence in setting up regular solitude experiences for others.

SETTING UP SOLITUDE

⊕ Pull the group together before departing and have them all hold their hands up like they are grasping a curtain, then everyone can lower the "veil of silence" with you. (Be sure to raise the veil again at the end.)

⊕ Explain that it is usually important to sit up during these experiences so they will be in more of an active mode rather than a passive one.

⊕ Provide more structure for school groups (e.g., with Magic Spots, dropping them off and picking them up one by one, giving them a task to focus upon the first couple of times, preparing a special booklet in which they can record their observations).

⊕ Emphasize that these experiences are not primarily for reading or writing or meditating, but for getting in touch with the flow of life. Some time should be spent just taking everything in, just absorbing the moment. Suggest that they tune in to the subtleties (shapes and shades and songs), but caution them not to let the little voice dominate their experience.

Hey! Would you believe it? We have wound this up and finished our pyramid. That's right. It has been another long section, but an important one.

Please take one more look at those nine crucial components of earth education before we turn our attention to what you can do with all this. . . .

WHYS	WHATS	WAYS
preserving	understanding	structuring
nurturing	feeling	immersing
training	processing	relating

CHAPTER SIX

eARTH eDUCATION... bUILDING yOUR oWN pROGRAM

First of all, let's review what we mean by an educational program. As I pointed out in the opening chapter, there appears to be a lot of confusion over that term. An educational *program*, as opposed to a collection of activities, should be a non-random, cumulative sequence of learning experiences focused on specific outcomes. And almost by definition it should be synergistic, that is, the whole should be more than the sum of its parts.

In the introduction to <u>Sunship Earth</u> I suggested that our field has produced lots of candy to choose from, but precious little in the way of programs. I suspect most people confused the means with the ends. They thought all that candy displayed for them at the workshops and conferences was the goal instead of a possible vehicle for achieving the goal. Of course, their confusion may be understandable when you realize that some of the folks producing the candy failed to point out to them what they were getting. Instead, they often handed out the bags!

The OBIS materials (for Outdoor Biology Instructional Strategies) serve as a good example of the problem. In that collection you find almost 100 activities, practically none of which deal directly and forcibly with the key ecological understandings needed in genuine environmental education (like energy flow and matter cycling). Instead, there are lots of activities dealing with secondary concepts like adaptations and populations (probably because they are easier to convey). Practically none of these activities include components designed to help the learners internalize the understandings in their own lives (unless you include the ubiquitous directive "discuss"), and almost none of them ask the participants to begin making changes in their personal lifestyles.

That's okay. OBIS didn't really claim to accomplish those goals, but neither did they hesitate to claim that they were doing environmental education either. They even thumbed their scientific noses at the rest of us by saying in their leader's manual that they didn't believe in the "sniff and appreciate" approach. In fact, they seemed to ride the wave of interest in EE with some glee. Untold thousands of teachers and youth leaders were led to believe that if they had included a couple of OBIS activities in their plans, then they had "done" environmental education. No one ever told them that

Sunship Earth awakens the feelings and imparts the understandings that will help young people better fulfill their responsibilities as crew members and passengers of this wondrous vessel of life that we share. For 4-5 days both leaders and learners embark on a re-training mission, getting to know their place in space and how they can go about keeping it healthy.

if they were really serious about EE, then it just wouldn't work to sprinkle some of these activities around like so many educational jelly beans.

Folks, bagging up a bunch of the candy that has been produced in our field and calling it a program is like tape-recording a batch of unrelated sounds and calling them a symphony. In earth education we aim to make each program a genuine, carefully-crafted symphony on living more harmoniously and joyously with the earth. That doesn't mean our programs are bereft of candy. It's what you do with the candy that counts. Perhaps the artfully arranged interior of a box of candy would provide a better image for what we should be doing, for a genuine educational program contains related pieces that are organized sequentially to produce a specific, cumulative effect. In a programmatic container everything is carefully ordered to support and enhance the desired outcome.

Such programs already exist in the educational community as a whole (there are some good ones out there in other fields like reading, math, etc.), so why do we suggest in our area that a program is little more than a couple of unrelated activities and a discussion (plus the obligatory poster in the hallway)? It's a mystery to me.

Make no mistake. Program building is tough work. It just won't do to bring some teachers together, toss some money at them and expect them to create good programs (not very often anyway). That is one of the worst mistakes we ever made in education. Designing a good instructional program is not the same as teaching one. You can be an excellent teacher and a terrible designer. They involve different skills. Most teachers have been trained in the areas of content and delivery, not in design and development. So please remember, just because you are a good learning helper doesn't necessarily mean you are a good program builder as well. For example, I am a much better builder personally than a helper. I tend to be too intense for lots of younger children (and some college students too!). I intimidate instead of motivate. But the important thing is that I realized this from the beginning and formed a team of good learning helpers to assist me in piloting the programs I designed. Who was the "Jimmy" in chapter two that I hid up in the trees at the edge of the marsh and watched twenty some years ago? He was a fourteen year old natural — a learning helper with the marvelous ability to put kids at ease, to coordinate and direct their energies without dampening their spirits. He still is. And we still work together today.

In this chapter we are looking at some tools to use in building an earth education program. In previous chapters we have erected the framework for the structure. We hope you will be able to use that framework and these tools in constructing a good program for your own setting and situation. However, if it doesn't seem to come together very well for you, please don't give up. Start looking around for some assistance.

Today, in The Institute for Earth Education, small, local branches are forming in several countries to do just that, to provide a place where helpers and builders can work together in developing and implementing earth education programs. One of the most successful of these branches grew out of a workshop I conducted several years ago on the south coast of England. Instead of going their separate ways at the close of the weekend, with each of them trying to operate in relative isolation somewhere, some of the participants decided to work together in building and implementing a genuine earth education program of their own. They rented out the regional field centre for part of a week during the summer, put their program together and sold it to the local youngsters and their parents as a holiday experience. And it is still going strong. Although these folks worked at different schools and centres (some held entirely different jobs), they decided to pool their talents in a local branch and get on with the job of actually building an earth education program. And in doing so they set the standard for our branches around the world.

Most importantly, please keep in mind that this group maintained a clear focus on what they were trying to accomplish. They knew the difference between earth education and environmental education, and they were determined to pursue the former. I am sure their success could be traced in large measure to the clarity of their vision and their resolute pursuit of it.

GETTING STARTED...

What about those who contend that supplemental activities are good because they give the teachers something to do, that a couple of activities just serve to get them started? This is what the apologists for supplemental activities always claim, but the reality appears to be that most teachers and leaders never do anything else at all. And why should they? The supplementalists provide them with no guidance, no models, no encouragement to go further. So don't buy this argument. It is often a cover-up for taking the easy way out. If you care about the earth, please help us get people beyond the supplemental approach. Tell them right off that you are going to work with them to build a complete program (that's what you are _starting_ on). Explain how you are setting off together on an exciting new path, and that you will be there to help them along the way. Remember: anyone can build an earth education program — anytime, anywhere.

ENVIRONMENTAL EDUCATION
VS. EARTH EDUCATION

ENVIRONMENTAL EDUCATION (TENDENCIES)	EARTH EDUCATION (AIMS)
⊕ supplemental and random	⊕ integral and programmatic
⊕ classroom based	⊕ natural world based
⊕ issues oriented	⊕ lifestyle oriented
⊕ focuses mainly on developing secondary concepts and conducting environmental studies projects	⊕ focuses largely on developing "ecological feeling" based on a combination of mental and physical engagement with the natural world
⊕ activity based	⊕ outcome based
⊕ claims to teach how to think, not what to think	⊕ claims to instill values and change habits
⊕ relies heavily upon conducting group discussions to achieve its instructional objectives	⊕ relies primarily upon participatory educational adventures to achieve its instructional objectives
⊕ integrates the inputs (messages) and consolidates the applications (projects)	⊕ consolidates the inputs (messages) and integrates the applications (projects)
⊕ infused with "cornucopian" management messages and views	⊕ infused with the ideals of deep ecology
⊕ accepts a wide range of definitions and intentions	⊕ rejects becoming everything to everyone

Please note the use of the qualifying terms in the headings above, i.e., environmental education tends to be and earth education aims to be. We realize that there are exceptions to these characteristics on both sides of the chart. However, in general we think you will find they represent accurate descriptions of the two movements.

Okay, if we haven't sounded too discouraging so far and scared you off, it is time to look at how you can go about building a program in your own setting and situation. We recommend a seven step process in designing an earth education program.

PROGRAM-BUILDING CHECKLIST

☑

☐ 1. Design Criteria

☐ 2. Rationale and Purpose

☐ 3. Goals and Objectives

☐ 4. Hookers, Organizers, and Immersers

☐ 5. Vehicles and Activities

☐ 6. 3R's: Reward, Reinforce, Relate

☐ 7. Transfer Components

"SEVEN STEPS in BUILDING a PROGRAM"

The first thing you must do is set down your *design criteria*. By that I mean such things as who are your learners (and what kind of background do they have), what is your setting (where will this program take place), who are your leaders, how much time do you have (an hour, a day, a week), what finances are available, etc. These criteria will be vitally important for everything else you do, so you will be wise to spend some time sorting out the specific details before you do anything else. (Be thorough.) I like to write the results up in the corner of a large chalkboard or list them on a poster to tack up in the corner of a "working" wall, for everything else must be matched up with these criteria to make sure you are staying on track. These are the limits within which you must operate. You will have to train yourself to habitually refer back to your design criteria as you work your way through the other six steps.

BE CAREFUL...

Don't let the criteria become the program at this point. A site survey revealing a pond does not mean that pond dipping automatically becomes a part of your program. This is a common planning pitfall that you should do everything possible to avoid.

Second, you need to explain your *rationale and purpose*. Why are you doing this program? What is the justification for its existence? Why is it needed? Who will benefit? In this case, your rationale provides the overall reasons, and your purpose explains the primary objects for which you exist. Since this is an earth education program, we have already done most of that job for you, but it would be good to go back and review the WHYS of earth education. That's what we do ourselves. We go back and check to make sure we are still doing what we said we were going to do in the beginning, and that our reasons for doing it have remained the same.

Third, set down your *goals and objectives*. Once again, we have spelled out the overall goal for you, i.e., living more harmoniously and joyously with the earth and all its life. But you will need to decide what specific part of that task you can hope to accomplish given your design criteria. Will you, for example, focus on building one key ecological concept, or on enhancing feelings for a natural community? In other words, you cannot build a harmonious, joyous relationship with the earth in a single day, so what can you do given your setting and situation?

This doesn't mean that this is all you will do, but it is where you will start. You see, there is no such thing as a one-day program in this field. Just as you would not call a one-day experience on the tennis courts a program in learning how to play tennis, you should not call a one-day experience at an outdoor centre a program in learning how to live more lightly on the earth. It is going to take a lot more work than that, and we should be upfront with everyone about this. In fact, most earth education programs will begin with an intense "springboard" experience in a natural setting, then continue back at school and home for some time.

While goals are usually worded for the leaders, objectives are worded for the learners. Consequently, in the process of pinning down your objectives, you will need to boil away as much of the verbiage as possible. As we have seen, one of the reasons environmental education got into so much trouble is that it kept trying to qualify every statement with a string of modifiers. Don't make the same mistake here. Get your statements worded succinctly. For example, the overall learning outcome for the Earth Caretakers program reads, "The earth, a wondrous vessel of life powered by the energy of the sun, needs our help." And in the Earthkeepers program we have boiled our objectives down to four simple sentences: "All living things on earth are connected;" "Getting in touch with the earth is a good feeling;" "Your actions on the earth make a difference;" and, "Helping others improve their relationship with the earth is an urgent task."

Let's start making this process more concrete ourselves right now. Let's say you have a group of 10-11 year olds coming out to your centre for a half-day visit. What could you really hope to achieve in that timeframe given the overall goal of earth education, and how could you set it up as a springboard experience?

You know, if I had a group like that coming out to a typical outdoor centre, I think I would work on something as simple as building a sense of wonder and a sense of place. Why? Because I feel such places often represent oases in the midst of a human-centered, human-dominated landscape. Oases where the overall systems of life, along with many of the other passengers who share the earth with us, can still be experienced. Oases where people can be immersed in the real stuff of life here, the natural communities and systems of the earth. And in a half-day I would just like to hook the kids on places like that — places that are filled with wild and weird and wonderful things that share the earth with us. I would try to grab onto them so strongly that they would go back and bug their parents and teachers and friends to go to places like that again. If I could accomplish that in a half-day visit to a site, it would be a great springboard for what I could do later back at school.

Somewhere along the way (you are never quite sure where this will happen in the design process), you will also need to come up with a *hooker*, an *organizer*, and an *immerser*. That's the fourth step.

The *hooker* is the motivator. It is the thing that pulls the learners in and motivates them to work on what you want them to learn. Remember, in Acclimatization we always said we wanted to pull the learners, not push them. Well, the hooker is the initial activity that pulls them in. You know yourself that motivating your

<u>Earthkeepers</u> is a magical, 2½ day, upper elementary adventure that captivates learners and leaders alike. Participants earn two keys for living in harmony with the earth (K = Knowledge, E = Experience) at a special Earthkeepers Training Centre, and two more after they return home (Y = Yourself, S = Sharing).

learners is half the task. If you can get them to want to do what you have set up for them, you are halfway home.

Our hooker for Sunship Earth is the message that the earth is in trouble and the learners are on a re-training mission so they can help save it. It is a message delivered by a mysterious voice in an important ceremony we call "Welcome Aboard." Imagine a special octagonal room of black cardboard walls set up inside another room. The kids enter this inner room with arms folded accompanied by slightly eerie, but earthy music. The wall in front of them contains a mural with a native American Chieftain sitting beside a stream in a forest glade. In the sky there is a mountain, and suspiciously, what appears to be a planet. There are large white "windows" on either side of this scene and one is labeled "Space," the other, "Time." As the music fades, a deep, resonant voice suddenly emanates from behind the mural:

> "Welcome. I hope you're enjoying your trip today. It's hard even to imagine, but at this very moment we are traveling at an incredible speed. This ship, which we call earth, is rotating to the east — that's to your right — at a speed of over one thousand miles an hour. At the same time, it is orbiting a medium-size star, the sun, at over sixty-five thousand miles per hour. Meanwhile, this solar system is whirling through the Milky Way galaxy and the galaxy is zipping through space, all at astounding speeds. I hope you're having a good ride."

The mysterious voice goes on to explain that there are two magic windows in the room through which the students can get a better grasp of where they really are. In the "window" labeled "Space" a hidden projector beams the film, "Cosmic Zoom," which takes them to the edge of the universe to see where they really live, followed a few minutes later in the "Time" window by scenes from the film, "Home," and the prophetic message of Chief Seattle. Afterwards, the voice from behind the mural explains what they will be doing at the Sunship Study Station:

> "Perhaps now you understand why you are here this week. You will be involved in many activities, and your purpose in all of them is to gain the feelings and understandings that Chief Seattle was

talking about. The key to the story of life on earth is contained in the letters you see on the side walls: EC-DC-IC-A. Take a close look at them. During your stay here you will learn the meaning behind the letters in that formula. You are to learn how the sunship operates so that you can act on that understanding before it is too late."

Each of our model earth education programs has a hooker like this one. In Earthkeepers, it is a visit to E.M.'s Lab, while in Earth Caretakers, it's a special package (including a riddle from a couple of Rangers) that is delivered to each classroom. These are all upper elementary programs, but we are working now on programs with their own integral hookers for both lower elementary and junior high school classes. Later, we will tackle both the preschool and adult levels.

The *organizer* is the thing that helps the learners keep track of what is happening to them in a program. It is the device that provides some logical way of holding onto the various parts of the experience, and if it is successful, serves as an accessing tool for the learners in the future. Don't let this one throw you. There is nothing unusual or esoteric or "American" about this idea. In fact, there has probably never been a nation of people on the earth who did not have organizers. That's what creation stories are. They are the largest organizers of all because they organize the fabric of people's lives.

Myths, legends, tales — these are all very large organizers. But there are also very small ones as well. What about some of those devices you used in school to remember lists of things — like the word HOMES for the names of the Great Lakes (<u>H</u>uron, <u>O</u>ntario, <u>M</u>ichigan, <u>E</u>rie, <u>S</u>uperior), and "<u>R</u>oy <u>G</u>. <u>B</u>iv" or "<u>R</u>ichard <u>of</u> <u>Y</u>ork <u>G</u>ave <u>B</u>attle <u>in</u> <u>V</u>ain" for the colors of the rainbow (Red, Orange, Yellow, Green, Blue, Indigo, Violet)? They are merely simple organizers. You probably still use some of them today. How about, "Thirty days hath September, April, June and November . . . ?" That's a common organizer for holding onto the number of days in each month. Of course, rhyming always makes a good organizer, and songs are even better. In fact, it is not uncommon these days to hear an exasperated teacher say, "These kids can't remember anything I tell them, but they remember all the words to those darn songs!" No

EARTH CARETAKERS

<u>Earth Caretakers</u> is an upper elementary springboard experience that takes only 1 day to initiate. It all begins when the participants receive a special package from the "Rangers"....

wonder, songs are some of our most powerful organizers. They not only have inherent rhythm, they often have a contextual storyline and vivid imagery to go along with it. I like to tell frustrated leaders, "Hey, if you would sing your lessons to them, they would probably remember those points just as well."

The *immerser* is the technique or activity that gets our participants over those common barriers that most people have erected between themselves and the natural world. Sometimes physical, often psychological, those barriers prevent the participants from grabbing on with both hands and just reveling in their contact with the flow of life. The immerser is an intense, perception-changing experience.

As we noted in the section on Immersing, to immerse means to plunge in, and that's exactly the initial feeling we want to convey — the feeling of diving into the flow of life in a natural area and wallowing joyously in its currents. It means casting off your artificial cloak of separateness for a while and making contact with the juices of life around you.

Getting that job done may involve something as simple as a bit of off-trail, cross-country rambling, or going out when everyone else is going in, like at night or on a rainy day. And yes, it may mean wading into a marsh or pond, or hugging the trees and kissing the flowers. Most of all though, it is demonstrating each day in all of your activities that it's not only okay to get wet and dirty, it can be downright pleasurable.

A good immerser contributes to each of the "Feelings" that we examined previously in the WHATS and draws upon the "Immersing" ideas that we explored in the WAYS, but it transcends all of them as well. Try not to make the pervasiveness of this idea seem too weird. There are plenty of examples around . . . Walt Disney achieved this effect in fantasy when he created the Magic Kingdom. A more familiar example would be the annual transformation of many homes as the December holidays approach, and lots of people have experienced getting caught up and carried along in a carnival parade like the Mardi Gras. All of these are artificial immersers of sorts, where the sights and sounds and smells combine to engulf the senses and delight the child within us.

TO KNOW IS
NOT TO FEEL...

On one hand we have designed fairly mechanistic and linear educational programs in our work, but on the other we are trying to convey a sense of the seamless fabric of existence — the undulating, ever-changing patterns of life as it continually unfolds. Some intellectualizing will no doubt aid our awareness of this, but it is at the intuitive, experiential level where we will most likely grasp our oneness with the flow of life.

In some situations, an immerser may involve encouraging people to remove some of the clothing they use to shield themselves from the elements, and it will certainly include getting them away from the eye-level approach to everything, then using all of their senses to absorb the area around them. Just be sure to temper such light-hearted, exuberant approaches with quiet moments of contemplation and awe.

However, a good immerser requires something more than contact. You have to lose your self-identity within a larger natural community, and join the rhythmical dance of energy around you. This is more difficult to achieve and almost impossible to describe.

In some ways it is experiencing intuitively our relative insignificance in the panoply of life spread over the earth. It's those moments when you feel most fully alive, yet lose your consciousness of time and space — when you seem to melt into your surroundings (often accompanied by a delicious little shiver upon your return).

Once you have known such moments yourself, it is easier to guide others into experiencing them, but overall it's a subtle combination of timing and suggestiveness that will require much practice and sensitivity to pull off on a regular basis.

In some settings you may be able to accomplish the intensified experience necessary for our immersion in the natural world through lots of free time or during program time exploration (perhaps using some of the natural awareness exercises from <u>Acclimatizing</u> enroute). In other areas, you will need to program everything more carefully, perhaps even scheduling immersers of varying degrees according to the background of your participants.

Of course, you can use our "vehicles" that focus on the feelings (like the Earthwalks and Immersing Experiences), but in that case you must make sure to select only those activities and settings that will assure the full-body contact necessary for the participants to really feel themselves merging with the elements and processes in a natural area. And please keep in mind that you cannot force those feelings. You have to pull, not push. Naturally, it may take more than one immersion to convince some of your participants that

"The rounded world is fair to see, Nine times folded in mystery...Though baffled seers cannot impart, The secret of its laboring heart, Throb thine with Nature's throbbing breast, And all is clear from east to west."

— Ralph Waldo Emerson

UEXPLORED
← ⸱·· TERRITORY →

HOBBITS
AND MAPS...

Hobbits were (or are)
little people who lived in very
comfortable holes in the
ground long ago. "They are
inclined to be fat in the
stomach; they dress in bright
colours (chiefly green and
yellow); wear no shoes,
because their feet grow
natural leathery soles and
thick warm brown hair like
the stuff on their heads
(which is curly); have long
clever brown fingers,
good-natured faces, and
laugh deep fruity laughs..."
In The Hobbit (or "There
and Back Again"), from
which this description is
taken, J.R.R. Tolkien tells a
tale about a remarkable
adventure that befell one of
these enchanting characters,
an adventure that began
with a strange map...

their lives will be enriched by such experiences, but
don't give up. The results of all your enticements and
cajoling will be worth the extra effort.

Let's get back to our example of a half-day outing.
If I had a group of youngsters coming out to an outdoor
centre for a short visit, I think I would make up a
special map of the place and send it to them a couple
of weeks ahead of time for my "hooker." But I wouldn't
make it a topographically detailed map; I would put
together a hobbit-like map. Do you know what I mean?
I would make it one that includes all those special
features and touches that add magic to a place.

In fact, if I had a centre, the first thing I would
do is start looking around to see where I could put up
a swinging bridge. There's a powerful hooker for you.
And sometimes it is just a matter of what you call
these things. Instead of referring to a physical feature
of the site as the old gully, call it the "Gulch of No
Return." Here is another item that has a lot of pull: if
you have anything at all that can be labeled as "Ruins,"
be sure to include that word on your map. And whatever
you do, don't forget to leave one whole area marked
simply, "Unexplored Territory." That'll grab them.

A few years ago, at a centre where I was conducting
a workshop, I suggested this idea of using a map for a
hooker. As it turned out, I went back for another visit
later and saw the finished product. Unfortunately, their
mapmaker didn't quite get the idea though because
instead of marking some of the surrounding areas as
"unexplored," he simply designated it as "wasteland."
It doesn't have quite the same appeal does it? (And
what did it say about how we view the wild areas of
the earth?)

gATEWAY to the
lost VALLEY

You should also include some drawings on your
map of some of the other "passengers" that the learners
will likely get to meet; some of those wild and weird
and wonderful creatures that live in your oasis. That
doesn't necessarily mean just the big things either —

in the U.S. people always want to put in the deer — because you can't usually count on them since they move around so much. But there are some things out there that you can count on. In fact, they are some of the most bizarre creatures on earth — the insects. That's right; just don't draw little pictures of them though, blow them up to emphasize their unusual characteristics.

Okay. That's enough. You are probably way ahead of me by now anyway. We are going to roll up our map and mail it to the class that is coming for a visit. But please don't send something like this to the teachers, folks. Don't waste it on them! Send it to the kids. What you should do is include in your application for a visit a space for the name of someone in that class who might need a little extra "hook." For example, maybe Billy thinks he is pretty cool these days, so guess who you should send the map to? Can't you just see him sitting there when the clerk knocks on the door saying there is a mailing tube for Billy? So Mr. Cool opens up the tube and pulls out your map of where the group will be going in a couple of weeks, then posts it for all to see. Oooowheeeee!

As I pointed out earlier, what a lot of kids are doing in school these days is what I call "S and S" time. So let's give them something to sit and stare at — a map of the special oasis of life they will be exploring. (We'll come back to the organizer and immerser later.)

The fifth step in building an earth education program is to look over our *vehicles* and *activities* to see what we have available to help you meet the objectives you have identified.

Vehicles are not activities. They are what we call our categories of activities. We use the term vehicle to describe these categories because that is how we are going to get to our goal. Each vehicle has a specific set of criteria to meet; it is designed to do a certain educational job, be it building concepts, instilling feelings, or enhancing solitude. By looking at the criteria you can find vehicles that fit your objectives, instead of taking the common approach of choosing an activity because you like it, regardless of how well it matches up with what you are trying to accomplish, then attempting to talk it into doing the necessary job. In earth education we start the other way round. And each

LURE OF THE UNEXPLORED...

I was so lucky when I was a kid because we still had textbooks then with areas of the world labeled "unknown." I can still remember sitting in class looking at maps like that and wondering what could possibly be in such places. You know, we should all get together and agree to designate some parts of our countries forevermore as unknown, just so people can feel that lost pull of adventure.

FINDING THE RIGHT ACTIVITY...

Since we are constantly adding to our list of activities, the best place to look over what we have available currently is in a separate institute publication titled, "Programs, Vehicles, and Activities," which cross-references all of our materials. By looking through this listing you can get an overview of your possible choices, then go to the source of the appropriate vehicles in order to make a more in-depth examination.

of our vehicles, like the Earthwalks, Concept Paths, Immersing Experiences, Conceptual Encounters, etc., may have several different activities in them. For example, there are over 75 activities to choose from (based upon your design criteria) for building just one Earthwalk that requires only 4-6 activities.

eARTH eDUCATION VEHICLES

Concept Paths . . .
"Adventuresome trails of ecological concept-building stations." Assimilating and applying to the natural world key ecological concepts (going from the abstract to the concrete and using activities between stations that encourage natural observation and examination).

Small group: 6-7 per station
Ages: 10-12

Discovery Parties . . .
"Guided exploration and rambling." Building a sense of wonder and place (using tasks that encourage making personal finds).

Small or large group: 5-15
Ages: 10-12[+]

Disguise-removing Techniques . . .
"Ways for uncovering the hidden nature of common objects." Using eye-opening and thought-provoking aids to reveal how everyday things are connected to the natural world.

Solo
Ages: 10-adult

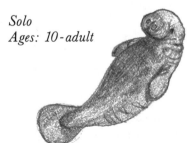

Earth Journeys . . .
"Extended experiences in living more harmoniously and joyously with the natural world." Building a sense of relationship with wild places (as an alternative to outdoor experiences that pit people against nature).

Small group: 8-10 maximum
Ages: teenage-adult

Solitude-Enhancing Activities . . .

"One on one experiences in natural settings." Being alone in the natural world, in touch with both the flow of life and with yourself.

Solo to large group
Ages: 10-adult

Environmental Habits Tasks . . .

"Practice in breaking bad environmental habits and forming good ones." Opportunities for examining one's lifestyle and crafting a more harmonious alternative.

Solo to large group
Ages: 10-adult

Immersing Experiences . . .

"Unusual opportunities for closeup contact and involvement with the natural world." Changing perspectives and breaking personal barriers (using multisensory activities with built-in sharing and self-expression).

Small or large group: 5-15
Ages: 10-adult*

Conceptual Encounters . . .

"Adventures in learning ecological principles." Developing a deeper understanding of one ecological concept at a time (going from the concrete to the abstract within a problem-solving storyline).

Large group: 15-20 ideal
Ages: 10-12+

Interpretive Loops . . .

"Awe-inspiring introductions to natural communities." Barrier-breaking sensory and conceptual experiences for getting to know natural areas.

Small group: 8-10 ideal
Ages: 10-12 ideal

Natural Awareness Exercises . . .

"Tasks for sharpening perceptual skills." Practicing the skill of becoming more fully aware and enhancing the ability of being more fully alive.

Solo to large group
Ages: 10-adult*

Values-Building Methods . . .
"Activities promoting personal reflection and growth." Developing and strengthening positive environmental attitudes and intentions.

Solo to large group
Ages: 10-adult

Earthwalks . . .
"Light, refreshing touches of nature." Sharpening senses while highlighting the richness of the natural world (using a series of special activities, usually from four to six, put together in a smooth flowing way).

Large group: 15+
Ages: 10-adult*

* *some individual activities suitable for younger children*
+ *activities for older participants are being developed*

Of course, you may find that we haven't created an activity that will match up with your design criteria. In that case, you are going to have to find one elsewhere or design and develop one yourself. We will come back to this situation a bit later, providing some general criteria for what makes up a good earth education activity. However, before we finish this section, here are some additional points you will want to take into account as you begin arranging the sequence of your activities:

⊕ Consider the participant's energy. Ecological concept-building activities command better attention in the morning; discovery-oriented experiences help energize the afternoons.

⊕ Full day and longer programs require time for frolic, exploration, and an extra dose of adventure. Check the "Discovery Parties" vehicle for ideas. Beyond that, "Time Out" recreational activities that convey a feeling of living in harmony with nature instead of having power over it may be used.

⊕ Watch out for "tradition traps." These include ceremonies or campfires that have always been run a certain way, environmentally questionable dining room procedures, etc. Earth education programs are unique in that the whole experience

is permeated by the overall goal. Everything from the organizing pattern to the living arrangements should support the effort. Be persistent. Question outdated policies. Emphasize the program. Even challenge your own perspectives.

⊕ Don't dilute it. Earth education should stand on its own as earth education. It's far too important to let it become watered down with other pursuits. Scientific investigations, elaborate socializing experiences, ropes courses, etc. will work best in programs that do a good job of getting their own objectives across.

⊕ Avoid putting activities together that use similar tools (such as a hand lens), techniques (such as matching shapes), or storylines (like the one about the rainbow collapsing).

THE POWER OF TREASURE...

When a cereal company in the U.S. announced that you could call a special number for a clue to the treasure map on their boxes, 24,000,000 telephone calls came in and they sold 18,000,000 boxes of cereal in six weeks.

In the example we have been looking at for the local outdoor centre, a Discovery Party is the vehicle that will probably best match up with our objective of building a sense of wonder and a sense of place. The Discovery Party was created to serve the need for activities that allowed for more free-wheeling exploration, where the participants could decide from moment to moment where to go and what to do. The idea was to have just enough structure to hold the group together and give them a sense of direction, but not so much that it would overwhelm them with the task. And building a sense of wonder and a sense of place are two of the major aims of a Discovery Party.

In this case, the particular Discovery Party that might best meet our need is the one in <u>Sunship Earth</u> called, "Lost Letters." The set up for this activity goes something like this: "Last summer I spent some time with an old man I met up in the mountains. He was not well — in fact, he probably didn't have long to live — but before I left he gave me this pouch. Inside, there are all these little bags of letters, and it appears there are clues or something in each one. I don't know what they mean. All I know is the old man said something about . . . treasure." Of course, a twinkle in the eye accompanies this explanation, but, as usual, it is just enough that the kids aren't sure if this incident in the mountains really happened or not.

At the beginning of Lost Letters, the leader hands one of nine small bags of cut-out letters to one of the kids, along with a task card about how to earn that clue for the treasure. The task accompanying each letter calls for finding three things that begin with the same letter in the alphabet, while using a behavior that begins with that letter too. For example, to earn their letter "I" each participant might have to find things icky, itchy, and inky, but all will have to be just an "inch" away. The learner "in charge" of each bag then verifies the discoveries and passes out the letters as they are earned.

Actually, this activity might also make a good "organizer" and "immerser" for our hypothetical visit to an outdoor centre. The participants could earn their letters as they explored that oasis of life. For example, they could work on the "D" over by the swinging bridge; the letter "O" might be earned when they were crawling through the stream; the letter "S" could be picked up in the unexplored territory, and so on. Just be sure to get off the trails and pause now and then between "letters" for some quiet immersion too.

In the end the kids will have earned nine letters during their exploration of the oasis, so they can sit down somewhere and figure out the meaning of their anagram. It turns out to be the word D-I-S-C-O-V-E-R-Y, and in the very bottom of the old man's pouch they find one last note:

> *"Congratulations. You have not only rediscovered the true meaning of discovery, but in the process you have discovered the real treasures of the earth."*

The sixth part of our program building approach focuses on the *3R's: Reward, Reinforce, Relate*. First we go back and look for pay-offs — for ways to reward the participants for completing the initial steps in the program. Remember, we view these as springboard experiences so we have purposefully made them fairly dynamic. We want to catapult a bunch of enthusiastic, focused learners back into their school and home settings to continue with the action. (We realize every bit of outdoor learning doesn't need some extrinsic motivation, but earth education programs, by design, are more intense in the beginning because they will have to be longer lasting.)

In the Sunship Earth program the participants place their stick-on letters in a circle on the covers of their Log Books, and we give them a special sticker of the earth to place in the middle of their discovery circle. (You could do the same, using either our generic Log Book or one of your own making.)

By the same token, we generate ideas in this step for how we can reinforce the key points throughout the experience. We want them to live what they learn and learn what they live. So we examine every part of the day, looking for those kinds of opportunities as well. Of course, this will also help us make sure we have begun relating these ideas and insights to their own lives at the same time. And that is the real trick, to make sure we build the mental bridges necessary for the learners to see what all this has to do with their daily activities.

And that leads us to the seventh and final step in our design process, what we refer to as the *transfer components*. If we don't build in this continuation of the action back at school and home, then we are just spitting in the wind. What really counts is what the learners are going to do with their experiences once they return to their usual settings. Don't forget: the point of earth education is change. If there's no change, there's no point.

You may have noticed by now, I don't much like the term follow-up. Over the years it has been my observation that those items seldom get acted upon. Oh, the term looks good on paper, but in reality it has become a common euphemism for not doing much of anything at all. In earth education we like to say follow-through instead, because our programs never end with one short visit.

Here are some points to consider as you work on the follow-through portion of your own program. . . .

CHANGING HABITS...

In earth education we have to change environmental habits. That should be the litmus test of every earth education program, does it aim to change specific environmental habits and how well does it succeed? For ideas, see the GUIDES FOR ACTION on page 162 and CHOOSING HABITS on page 146.

CREATING tRANSFER aCTIVITIES

☛

☐ Start with personal habits.
  Deal with immediate and manageable individual behaviors, and begin with the easiest things to change in order to provide confidence for tackling the harder ones later.

☐ Think in terms of energy and materials.
  (How do we get them, what do we do with them, what are our alternatives?)

☐ Focus on lifestyle choices.
If people of any age can describe the way of life they seek, then they will have something to shoot for that will make all the personal changes easier. Remember: earth education programs end at home.

☐ Emphasize action.
Avoid a lot of lengthy research or discussion or reports. Stress the "how-to's" so each participant has something personally to do about any problem.

☐ Look for ways to work together.
Structure your experiences so people can support one another and share their personal successes.

In the example we have been working with here, when the kids arrive we can take their poster-size map that we sent them and give them small, individual maps to follow instead. Then while they are off exploring, we can turn the large map over and mark off on it some of the key features of their own neighborhood — the school, a couple of main streets, the vacant lot, etc. When the kids return from "Lost Letters," we can say, "Hey, you've explored this oasis, but there are lots of pieces of places just like this back in your own neighborhoods. Some of them may be just little pocket oases, others mere remnants of what they once were, but such places are still out there waiting to be discovered. So you can continue your explorations when you return."

Of course, the kids already know about such places. Some of them represent special spots that they have been familiar with for a long time. In reality, they probably haven't been very kind to such places either, particularly the boys. I don't know what causes it, but the male of our species seems to go through this strange phase — something comes over us for several years — when we have this overwhelming desire to take things apart, literally to bust things up. Sadly, a lot of adolescent boys are out there every day doing just that, flailing away at the other life of the earth. I call them earth bashers. However, this might be our chance to turn them around, to change them into earth champions instead. Just think what we could accomplish if we could get the kids to look out for the other things living in their neighborhoods, to monitor the health of some of these pocket oases of life. Just imagine a whole wall of a classroom turned into a "monitoring" map of the local area. (Yes, with some places marked as "ruins," and others as "unexplored territory.") And picture color-coded information tags or discovery announcements hanging from those spots where various "champions" are watching out for the other life that shares the planet with us. In short, our "springboard experience" at the local outdoor centre or park could get the learners started on the important task of examining their own behavior in relation to the communities and systems of the earth.

Naturally, the broad brush strokes we have been using in our example here represent only the barest sketch of a program. It would take a lot of work to fill in this picture with the understanding, feeling, and processing details required for composing an adequate representation of an entire earth education program, but I hope this has given you a grasp of the process. (In the next box, we'll recap some of its key points.)

tEN CHARACTERISTICS OF aN eARTH eDUCATION pROGRAM

An earth education program:

1. Hooks and pulls the learners in with magical experiences that promise discovery and adventure (the hooker).

2. Proceeds in an organized way to a definite outcome that the learners can identify beforehand and rewards them when they reach it (the organizer).

3. Focuses on building good feelings for the earth and its life through lots of rich, firsthand contact (the immerser).

4. Emphasizes major ecological understandings (at least four must be included: energy flow, cycling, interrelationships, change).

5. Gets the descriptions of natural processes and places into the concrete through tasks that are both "hands-on" and "minds-on."

6. Uses good learning techniques in building focused, sequential, cumulative experiences that start where the learners are mentally and end with lots of reinforcement for their new understandings.

7. Avoids the labeling and quizzing approach in favor of the full participation that comes with more sharing and doing.

8. Provides immediate application of its messages in the natural world and later in the human community.

9. Pays attention to the details in every aspect of the learning situation.

10. Transfers the learning by completing the action back at school and home in specific lifestyle tasks designed for personal behavioral change.

Perhaps you have been thinking that all this is fine for a sense of wonder and place, but what if someone wants to work on building some ecological concepts during a half-day experience instead of the feelings. What would you do then? Before we can answer that we need to pull back and look at our whole approach to concept-building in more depth.

"dEVELOPING mENTAL fILING fOLDERS"

Sometimes when I start talking about concepts in our workshops, I can almost hear people out there mentally groaning, "Oh, no. Here comes all the heavy educational jargon." First of all, that's true, it does. But don't let that put you off. Concepts are like mental groupings. They are categorizers in the mind. And they come in all sizes, from the very small to the very large. Some of them represent fairly concrete notions like the chair or book you are probably using right now, while others are pretty abstract, like love and nature, two of the things that may have motivated you to read these words.

In earth education, we like to think of concepts as filing folders in the brain's filing system. Imagine that your mind is a gigantic warehouse full of filing cabinets. There are literally hundreds and hundreds of rows of them. In the past much of outdoor learning has been an attempt to label some of those drawers with categories like plants, animals, soil, water, etc. Next, the outdoor leaders would try to label for you each of the folders in a drawer with the name of an appropriate object in that category. A few outdoor school folks and amateur naturalists have always been fond of this approach, arguing that this way the learners could file away future bits of information on their own. That sounds fairly convincing on the surface, but sadly, they were usually describing what *they* liked to do instead of what their participants were *likely* to do. As I suggested in the section on naming and labeling, for many learners this simply guaranteed that they would end up with lots of empty filing folders. Think about yourself for a moment. Picture some of your own folders stored away up there. How many of them are relatively empty? Chances are the names on some of the tabs have even begun to fade, but the folders are still up there waiting on that long-promised day of fulfillment.

What we want to do with that system in earth education is fairly simple. We want to organize those folders in terms of the systems and communities of life. We want folders for energy flow and cycles and interrelationships. We want to build folders for a marsh and forest and desert.

Where do concepts come from in the first place? Obviously, we begin forming concepts before we have words for them. (Or we have the folder before we have anything written on the tab.) One of my favorite stories about concept-building is about a little girl I was watching one evening in a laundromat. You see, I like to go to a laundromat like I went to the marsh, i.e., I hide in the trees and watch. I have always felt a laundromat represents a real slice of life. You just sit back and watch the parade; eventually, everything happens there. Well, one night I am watching the parade from my perch in the corner when a small child spots a moth flopping and fluttering around on the floor. She immediately goes on the attack, running over and shouting, "Bird, momma, bird!" Momma, who was over shoving some things in a tank across the way, turned and replied rather sharply, "Oh, that's just a moth." Well, the little girl completely ignored that and continued dancing around in glee saying, "Bird, bird!" I think we can assume that that little girl's concept of bird was something with wings that flops and flutters, and at that moment the moth matched up with her notion of a bird. Eventually, she will have other experiences, gather new data, and probably end up with a

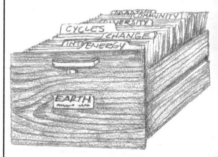

concept for bird, another for moth, a third for butterfly, and so forth. We can even hope that she will get a little better feedback than she was getting from Momma in this case, since she was ignoring it, but that is the way concepts often get formed, i.e., through concrete firsthand experience that captures the attention.

One of the myths about concept formation in our field is that a good way to form ecological concepts is to do a lot of analyzing and testing, a lot of quantifying and measuring. Untrue. In fact, those activities appear to represent a fairly sophisticated level of concept formation, not at all where we want to start.

In earth education we believe good concept building, particularly for youngsters, requires concrete, participatory experiences that deal directly with the concept, not with its ramifications. For a child riding down the highway in an automobile, a small brown and white animal in the distance on the hillside may well represent a dog. "Look, doggy!" After all, it is the right color and shape from the kid's perspective. You can repeat over and over that this particular animal is a cow, but you are likely to keep getting "doggy" in response. The idea of cow may be too abstract. But take that child out to a farm for just one close encounter with "Bessie" and guess what? You will have formed a cow folder, and since there are enough continual reinforcements for that concept in our lives, that filing folder will probably remain with the child throughout its years.

Before we get too far along in all this, there are a couple of dangers you should be aware of though when it comes to this concept-building process: misfiling and misforming.

The first is often the result of believing that all you have to do is to take the kids out there and point out to them a lot of examples of the concepts. Remember the "follow-me, gather-round" approach? "Okay, kids. Now here's an example of . . . yackety, yackety, yackety." After all, the reasoning goes, just like the story about the cow you can introduce the kids to the concepts in their natural setting. Unfortunately, ecological concepts don't represent things, but processes. And although you might be able to grab onto a piece of the process (like the water cycle, for example), it is very hard to get a handle on the whole this way. It's just too abstract.

As a result, a lot of these examples end up getting misfiled. So in earth education we want to build the filing folder first and fatten it up with some examples later. In fact, that is what we call our examples: filing folder fatteners.

I used to tell an embarrassing tale about my own experience with concept formation. Years ago (I won't tell you how old I was), I was with my parents in the south visiting a Civil War site. After the tour, the leader asked if there were any questions, and I replied, "Yes, where's the tunnel?" The ranger looked rather puzzled, while everyone in the family acted like I didn't belong with them. You see, I thought the underground railroad was. . . . Well, you will have to admit, if there was a tunnel that ran from Georgia to Canada, it would be worth seeing, right? I guess you could say my concept of the underground railroad had gotten off the track, and that is the second danger to watch out for in concept formation. The concepts get misformed. In fact, there is probably not a month that goes by that everyone out there doesn't have that experience where an old concept suddenly "clicks in" anew for them. You know what I mean, you are listening to someone, or watching television, or reading something, and in the back of your head there is this little voice saying, "Wow, I never knew that's how that worked." Chances are good that the concept was just too abstract for you, and that is why it got misformed in the beginning.

Our most important task then in this part of earth education is to get the concept into the concrete. Stop for a moment and examine the chair you are sitting in right now. (Feel how it supports you. Note how its areas of pressure and texture effect you.) Okay, you have a mental filing folder somewhere in your filing system for the concept of chair. Inside that folder there are all kinds of sizes and shapes and materials that go into making up your notion of chair. Right now, that concept is a very concrete one for you. Why? Because you are using it, and I have called your attention to it. But what would happen if you got up? In a sense your concept of chair would become a bit more abstract. You wouldn't be using the concept, and you wouldn't have much contact with it, so your concept of chair would be less concrete. What if you could only see a part of that chair, say from an adjoining room? In that case, it would become even more abstract for

THE UNDERGROUND RAILROAD...

For our overseas readers, in the decades preceding the American Civil War (1860-65) an elaborate underground system grew up for helping people escape the slave-holding south and reach the free states and provinces in the north. Hidden in the cellars and barns of abolitionists (often pictured in schoolbooks as someone with a lantern leading people down into something), the fugitive slaves were passed at night from one safe "station" to another by these anti-slavery "conductors." Over the years this route to freedom became known in the U.S. and Canada as the underground railroad.

you. In fact, if you brought a small child in and she could only see a part of a chair through the doorway, it might not be concrete enough to match up with her filing folder for chair. But since you have a fat filing folder for that concept, with numerous examples accumulated over many years, you can see just a small part of a chair and still match it up with the correct filing folder.

Next, what would happen if you left the room entirely, or turned around and faced a wall? Now you could not see a chair, you would not be using a chair, you would have no contact with a chair. The concept of chair for you would be an abstraction. In earth education our first and foremost task is to get the concept into the concrete. If this makes sense, let's look at how we go about doing it.

Traditionally, in the field of outdoor learning there have been two basic approaches to conveying ecological concepts. I call them the lecture method and the experimental method. For example, in the lecture method the leader takes the kids outside, holds up a leaf, and explains that photosynthesis is taking place in there . . . yackety, yackety, yackety. In the experimental method the leader has the kids wrap some leaves out there in tinfoil, then come back several days later and look inside to find that the leaves have died — "because they didn't get any sunlight . . . yackety, yackety, yackety."

However, in both of these cases I don't think the activity got photosynthesis into the concrete at all. In the first example, what was made concrete was not photosynthesis, but the leader and the leaf. And in the second case, I suppose what they made concrete was that if by chance you ever run across some tinfoil wrapped leaves out there somewhere, you will know they are probably dead.

It took me about five years to come up with an alternative to these traditional methods for conveying photosynthesis. And when it finally came to me what we should do instead, it was so simple that I couldn't understand why it had taken me so long to figure it out. It was easy. If we wanted to get photosynthesis into the concrete, then we should go in there where it was happening. And that's what we did. Today, in Sunship Earth we build a gigantic leaf on one of our

AN ESSENTIAL DIFFERENCE...

In earth education we give our learners the essence of a concept, rather than "drawing it out" by playing "Twenty Questions," then immerse it in direct experience. Afterwards, we ask the learners to do something with what they have just done, and then begin the on-going task of relating it to their own lives.

Concept Paths. The kids don their "chlorospy" hats and take their flashlights and crawl up the stem of the leaf to see what is going on in there. Inside, it is set up like the assembly line in a factory (that's why we call this station on our path the "Food Factory"), and the learners take their places on the line and try to manufacture some sugars. First, molecules of air (white ping-pong balls stuck together with velcro and labeled CO_2) come down these chutes from the ceiling. Next, molecules of water (blue ping-pong balls stuck together with velcro and labeled H_2O) come up the stem of the leaf. Then the kids on the assembly line take the molecules apart and try to put them back together to make a molecule of sugar, but the velcro doesn't match up so they can't do it. Something is clearly missing. That turns out to be the yellow ping-pong balls (labeled E for sunlight energy) that come popping through all these holes in the roof of the factory. Using the energy balls, the velcro finally matches up correctly and the students can put together a molecule of sugar. However, the O_2 ping-pong balls won't fit on anything now so they are poked out through the holes in the sides of the leaf as a byproduct of the process. Get the picture? It's not hard to figure out is it why this kind of hands-on, concrete, participatory activity can have a lot more focused impact than holding up a leaf and talking about the process inside, or setting up an experiment on what hasn't occurred.

At this point, I think it will help if we return to that analogy of the human brain as a gigantic warehouse full of filing cabinets. What we want to do in earth education is to label one of the drawers in one of those cabinets with the words, "Earth: Our Place in Space." Next, we want to pull that drawer open so we can put some filing folders in it about how life functions ecologically on this planet. If you have ever hidden in the trees and watched much outdoor learning underway, you know yourself that lots of times the leaders are doing their thing out there, but the learners don't even have the drawers open. We need something to reach out with and hook onto one of those drawers to pull it open for our learners. (Right, this is where our term "hooker" originated.)

After we get the learner's attention, we want to build some filing folders for the basic ecological concepts that govern life here and put them in the drawer we've labeled. Remember: we want to focus on the processes

of life instead of the pieces of life. So we want to build fewer, more fluid folders about the big picture of how life works here. And on the tab of each folder we want the name of the concept, plus a short, key concept statement that gets at the essence of what is in that folder. For example, "The building materials of life must be used over and over" for cycling, or "All living things interact with other things in their surroundings" for interrelationships. We think of these as our billboard messages, and aim for about ten words or less. (You can't read more than that when you are zipping down a highway.) Don't let this idea throw you either. There is nothing unusual about it. In fact, the people in our societies who pay the most attention to how folks learn use this idea every day. Logically, you would think that would be the educators. Wrong. Do you know who the people are that really pay the most attention to how folks learn? Right, the advertisers. And they use this idea all the time. Just look around and you will spot numerous examples of these essence statements in action. Recognize this one, "The World's Favorite Airline," or how about, "The Drink of a New Generation?"

Occasionally, someone will complain that we simplify too much in our work, that our "billboard" messages are misleading or inaccurate. For example, a science teacher at one conference objected to the key concept statement we use for energy flow, "The sun is the source of energy for all living things." He wanted us to change it to "most living things," and he couldn't seem to understand how that qualifier would make a real difference in the impact of the message. In fact, it would have been much better to change it to "The sun is the source of energy for life on earth," because that way it would still convey the overall importance of the sun's flow of energy here. Actually, we have stuck with the original statement because if and when the youngsters learn about those creatures deep beneath the seas that are living off of the energy from the earth's molten core, or those chemosynthetic bacteria, it just makes them all the more special, without detracting from the essential message that it is the sun that makes life possible here. Besides, you don't tell kids everything anyway. You generalize in order to make a larger point. We tell them the earth is round, don't we?

The real problem demonstrated in this interchange is how some leaders get so bogged down in the technical details that they fail to communicate essential messages. Take a look at these environmental education principles from one of those new state curriculum guides that have become all the rage in the U.S.:

> *"Earth's environment is a complex, interrelated, interactive, dynamic, constantly changing macrosystem called the ecosphere."*

> *"The characteristics of an ecosystem, derived from the interaction of its components, differ from the characteristics of individual components and can be understood only when studied as a complete functioning unit."*

Whew! Is it any wonder that leaders lose sight of what they really need to get across? I did get a chuckle in this case out of the presentation of that first principle in the guide, for right next to it in the margin was printed John Muir's classic quote, "When we try to pick out something by itself, we find it hitched to everything else in the universe." It was like the authors understood the importance and power of Muir's approach, but just couldn't bring themselves to be that elementary.

In <u>The Earth Speaks</u> I told the story about the university that brought in an astronomer and an actor to present lectures to two different groups on the same essential messages about the nature of space. And guess which of the classes retained the most and enjoyed it the best? Of course, it was the performer, for he knew about good communication, while the astronomer just knew about the stars. If we are serious about an educational response to our environmental crisis, then we must simplify our messages and prioritize their importance.

Back to the folders. In addition to the statement on the tab, I wanted to add a visual image to the front of each of our key concept folders. You see, most people are visual learners. They need to see something to help them make what you are talking about more concrete. It took several years, but we finally worked out the visual representations shown here for each of the four major concepts that earth education is based upon.

ENERGY FLOW

CYCLES

INTERRELATIONSHIPS

CHANGE

Together, these images combine in a powerful symbol that can be used both before and after our educational experiences. And its primitive simplicity speaks to our heart as well as our head. (In our upper elementary Earthkeepers program the learners see this symbol before their experience begins, but only find out that they are to gradually discover its meaning during their activities. Then as they work their way through our four Conceptual Encounters, they learn what each part represents.)

Finally, once we have built some filing folders about life on earth, we don't want to just stuff them in our drawer any old way. We want to organize them. You know yourself that if you can organize the filing folders in your own drawers at home it will be a lot easier to put things away, and much easier to find them again when you want them. In Sunship Earth we organize the folders in our drawer around the formula EC-DC-IC-A, which represents the story of all life on earth. It is a simple, rhythmic device that the learners can use to hold onto the ecological concepts in the program. (Once again, that's what we call the "organizer.")

EC-DC-IC-A

Energy Flow
Cycles
Diversity
Community
Interrelationships
Change
Adaptation

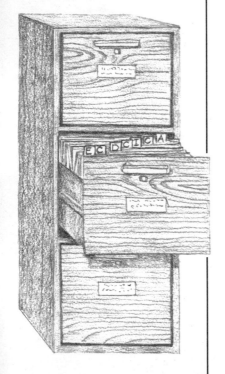

If you are thinking by this time that our analogy of filing drawers and folders is too mechanistic, then picture these concepts and characteristics as an arrangement of perceptual lenses through which you view the world. And when each of them has been properly developed (or cleansed perhaps) it is like looking at the world on one plane through the eye of a dragonfly — each facet representing one part of the understandings, feelings, and processings that make up an earth education perspective — yet all working together on another plane to form an overall composition of "the big picture" of our place in space. At any given moment then, you can see any scene before you as either a whole or as a multi-faceted, interlocking arrangement of concepts and characteristics.

All right. Let's go back to our seven step design process. What if you wanted to work on the understandings instead of the feelings during your half-day springboard experience? It would work the same way as before. You would look through our list of vehicles to select those that best met your program needs, then check out those specific concept activities that seemed most applicable to your setting and situation. In this case, you would have a choice of either putting together a Concept Path, like those in Sunship Earth, or using the Conceptual Encounters, particularly the ones in Earthkeepers.

We published the four Conceptual Encounters for Earthkeepers in a separate binder, but the Concept Path activities are only found in the <u>Sunship Earth</u> book. However, in this case, if you want to set up a Concept Path (without doing the whole Sunship Earth program), you may obtain from us the camera-ready masters you will need (along with the copyright permission and the Leader Cue Cards) to print your own concept-building booklets. It sounds complicated, but it is really fairly simple. Just read through <u>Sunship Earth</u>, pick out the activity stations that you would like to use from the Concept Paths, and contact the institute for the appropriate order forms. This will probably make more sense after we look a bit closer at the learning model Sunship Earth is based upon.

"tHE i-a-a lEARNING mODEL"

The basic idea here is that learning is . . . experiencing, responding, changing. Or, to put it more programmatically, we take something in, do something with it, then use it. So the "I" in our model represents the Informing or experiencing or taking in level of the learning. Of course, there are all kinds of ways we ordinarily do this in our own lives, and it is the same in formal education: we see, hear, read, etc.

Next, we have to do something with what we took in. We have to respond to it in some meaningful way. That's the first "A," for Assimilating. We repeat it; we try it out ourselves; we write it down; we fit it in with other things we have already learned; we practice, etc. In short, we process the information.

Finally, we are ready to put it to some use, to actually incorporate a change in our behavior or mental patterns. For if we don't really do anything with it (more frequently at the outset, every once in a while later on), then it will begin to fade rather rapidly. That is why the second "A" is so important. It is the Applying level of the learning.

Learning is one of the few things that lasts longer the more you use it, which may explain the estimate that we lose over 90% of what we learn in school. Other than playing trivia games or taking those quizzes in the magazines, there is little use for much of what we have so laboriously learned. Ironically, we have structured a society in which there is little we can actually do with much of our schooling. Until we solve that problem, there is no point in decrying the intellectual caliber of the students we are turning out.

The problem with most outdoor learning today is that the leaders confuse the role of the outside experience in the I-A-A learning model. They bring the kids out for applying when it should be for assimilating. Consequently, their work becomes a culminating experience instead of a springboard for what can take place back at home and school, where the applications could really count.

In earth education we use the I-A-A learning model in most of our programs. It keeps us from seeing our activities as ends in themselves, or as add-on's that have little relevance for the intended outcome. In short, our activities are integral parts of a sequence that begins with identifying what we want the participants to learn and ends with tasks for putting that learning to use.

It has been said by one educator that we do the right thing in earth education for the wrong reasons. Maybe so. All I can tell you is that I don't believe the source of our ideas can be attributed to any particular learning theory. Although we attempted to justify our educational approach in <u>Sunship Earth</u> by substantiating various elements of it using the views of the most widely recognized learning theorists (i.e., Skinner, Bruner, etc.), we made no claim to be following either a Skinnerian or Brunerian approach (nor any one of a half dozen others). And our generalized use of some of their insights or technical applications does not mean that we were unaware of other approaches. However, their work obviously influenced our thinking about the learning process.

Nor do we believe that all learning needs to use the same strategies. After one of my "Mission Gone Astray" speeches at a regional EE conference a few years ago, a local professor complained that I was not modeling the process I was advocating. But I was dealing with pre-motivated, pre-selected adult learners who had enjoyed considerable experience in the field and come voluntarily to listen to my comments. I was not working with inexperienced 10 year olds who had been forced to attend. That makes a world of difference. Frankly, I think this particular colleague's problem was that he didn't like my message and didn't know how to respond to it, so he attacked its delivery instead. A lot of leaders seem to fall into this trap. The secret is to select the best learning strategies for the setting and situation in which you find yourself.

We believe good learning in mass education is a product of good instruction, and it is our task as program builders to select the educational approach that is most suited to the outcomes we have in mind for the setting and situation within which we must operate. In earth education we are trying to take theoretic knowledge and put it into a concrete context for our learners. To do this we are telescoping lots of abstract understandings into intensified perceptual learning experiences.

While it may be true that people can gain considerable understanding through perceptual involvement alone (native people have always appeared to have a better grasp of ecological realities than so called modern folk), we don't have the luxury today of either those settings or situations. In most cases, we will have neither the time, nor the leadership, nor the location to be able to assure that our learners will gain enough desperately needed understandings about the earth's natural systems and communities through perceptual involvement alone. The institute's coordinator in Britain, Ian Duckworth, tells a delightful story about waiting on a field trip to see a food chain complete itself:

> *In my many years in the field with groups of young people, I have only once witnessed a food chain in action. It was such an event it is etched on my memory. In years to come my grandchildren will gather at my knee and ask, once again, to be told about the day I saw the food chain. So I'll tell it to you now. . . .*

> *I was out with a group on a local heath. At the time we were on our hands and knees looking closely at a small insect-eating plant. It was sticky to touch and they were fascinated, and so quiet that a damsel fly pitched near us and started to clean itself. They watched as it turned its head round with its feet dusting itself down. Suddenly — as we watched — a larger dragonfly pounced on it and commenced eating. They were spellbound. (I learned something too: the first thing it did was bite off the wings!) Well, we must have disturbed it because half way through its meal it took off carrying the remainder. We watched it zoom over the nearby bog pool . . . and as it passed by a bush, up jumped a bird and grabbed it, right before our eyes! We stood up open-mouthed — and waited — and waited. Well,*

*we gave the sparrowhawk about 10 minutes, but it
was late and we had to get back for our tea. . . .
You see, you can't rely on nature and chance to
teach the essential processes of life. After all, it took
the scientists many years to work them out, and we
have rather less time I'm afraid.*

In earth education then, we are attempting to bridge
the gap between theoretic and pragmatic knowledge
through an emphasis upon what we have termed ecological feeling. We don't think we can wait around for
that apocryphal food chain to complete itself (that's
the one that many outdoor leaders claim the kids get
to see on a traditional field trip). We are convinced
that we have to construct specific, structured learning
experiences that combine the understandings and the
feelings in a new way.

At this point, I think it will help to review our
design process before moving on to examine some
additional tools. By now, you should have a fair idea
of how you would proceed if you wanted to develop a
short "springboard" experience around either the concepts
or the feelings. Using the seven steps we have outlined,
you should be able to put together a solid half-day
experience:

1. establishing your design criteria
2. reviewing our rationale and purpose
3. determining appropriate objectives
4. creating your hooker, organizer and immerser
5. selecting our vehicles and activities
6. adding elements for the 3R's
7. developing your transfer components

But what would you do for pre-schoolers, or
teenagers, or adults? Or for a two day or two week
experience? It would work the same way. For instance,
although most of our examples in these pages have
been for programs for youngsters, there is no reason
why a program for adults couldn't be developed following
the exact same process. (Eventually, as the chart below
indicates, we plan to do it ourselves.)

ACHIEVING
RECOGNITION...

The Institute for Earth
Education has set up an
accreditation system for
those who develop their
own earth education
programs. If you would like
to get some positive
feedback on your work, or
share your efforts with
others, or have your
program recognized for its
quality, please contact the
institute for an application.

MODEL
EARTH EDUCATION PROGRAMS

Earthborn • ages 0-3 • a series of "earth bonding" experiences • idea stage

Earthlings • ages 4-5 • activities for building a sense of wonder, place and care • ready for piloting

Nature's Family • ages 6-7 • special events organized around the characteristics and needs of life • ready for piloting

Lost Treasures • ages 8-9 • classroom activities and field excursions focusing on natural communities • preparing the first pilot

Earth Caretakers • ages 10-11 • a 1 day trip away from school that initiates year-long learning and exploring • final piloting underway

Earthkeepers • ages 10-11 • a 2½ day experience for preparing to use less energy and materials • published

Sunship Earth • ages 10-11 • a week long adventure in discovering seven key ecological principles • published, piloting revisions

SUNSHIP III • ages 13-14 • a 2½ day experience emphasizing the choices to be made about our impact on this planet • final piloting underway

Earthways • ages 16-19 • a small group working together to develop personal responsibility while exploring the richness of the earth • almost ready for piloting

Earthbound • ages 20- • weekend experiences for building a more harmonious and joyous relationship with the natural world • idea stage

All of our programs also focus on helping the participants form some good environmental habits.

"CREATING hOOKERS, ORGANIZERS, AND IMMERSERS"

Don't forget you are going to have to come up with your own hooker and organizer somewhere in this process as well as an immerser. Finding a hooker and organizer that will work well together and fit flawlessly into the whole experience is one of the toughest parts of program building. Fortunately, the Immersers are usually a bit easier.

In Earthkeepers we came up with an integrated, double hooker and organizer. The mysterious character of E.M. in that program, and the secrets behind those initials, are both a hooker and an organizer. And E.M.'s KEYS both organize the components of the entire experience (K = Knowledge, E = Experience, Y = Yourself, and S = Sharing), and serve as a hooker because the kids know they are going to receive their own set of keys (to unlock E.M.'s boxes) as they go through the activities.

For our work in designing hookers and organizers I like to look at the early years of human growth as passing through three rough stages. Think of them as the three "I's" of development: imitative, imaginative, intellectual.

The first or imitative phase lasts until about the time youngsters lose their baby teeth. During these early years children are captivated by the actions of others and will often spend hours imitating their activities. Consequently, they are usually motivated by opportunities to engage in actions they have seen adults perform. (It's also obvious that they want to mentally organize things, or at least establish context, because they will frequently "talk" the objects they play with into some sort of storyline, regardless of how simple and commonplace those things may be.)

The second phase is an imaginative period where the learners discover and develop their ability to enter a world of fantasy. Their previous imitative actions can now take place in a far richer and more varied context, even though they often need concrete representations for key elements of their imagined worlds. In this stage, motivation is often triggered by the promise of larger-than-life experiences that will transport them somewhere else.

I know, some people object to our use of fantasy, saying that we are bringing the kids out to the natural world for other than the purest of reasons. But hey, we are competing these days with some pretty energy intense diversions. As leaders, we all know that any patch of natural land is chuck full of wild and weird and wonderful goings on, but the kids don't. We have to hook them first to get them out there. In earth education the focus of our activities remains on the natural things, but we add a little fantasy now and then to pull the learners in.

The third phase of development appears to begin at about the time of puberty. This is the intellectual phase. Now the life of the mind takes on a new importance. Young people relish the hours spent exchanging notions and insights with their classmates and friends, and abstract ideas captivate them as they are able to move further away from the concrete.

In brief, the family is of utmost importance to youngsters in the first phase, while their heroes capture their attention in the second, and their peers dominate them in the third. So when you are looking for ways to hook your learners and organize their experience, you should consider which of these phases you are addressing. (A small hint: Another way to remember the time span for the three "I's" is to think of them as roughly corresponding to the years represented by the designations pre-school, pre-puberty, and pre-adult.)

"The dynamic principle of fantasy is play, which belongs to the child, and as such it appears to be inconsistent with the principle of serious work. But without this playing with fantasy no creative work has ever yet come to birth."

— C.G. Jung

gENERATING pROGRAM iDEAS

Use these lists to trigger ideas. After you have worked with them separately for a while, try taking one item from each column and combining them for an overall storyline to be developed (e.g., sealed container, formula, going out at night without a light).

hOOKERS	ORGANIZERS	iMMERSERS
⊕ ceremonies	⊕ anagrams	⊕ going barefoot
⊕ exploration	⊕ riddles	⊕ rolling down a hill
⊕ challenge	⊕ clues	⊕ burying yourself in natural things
⊕ adventure	⊕ formulas	
⊕ parades	⊕ secrets	⊕ going out at night without a light
⊕ treasure	⊕ visual images	
⊕ mystery	⊕ symbols	⊕ walking in the rain
⊕ journeys	⊕ numbers	
⊕ sound and light	⊕ rhymes	⊕ hugging trees
⊕ special effects	⊕ songs	⊕ crawling
⊕ sealed books or containers	⊕ stories	⊕ using a blind-fold to heighten other senses
⊕ unique or unusual things or events	⊕ mnemonic devices	⊕ sleeping under the stars
⊕ setting records	⊕ acronyms	⊕ wading
⊕ very high or very low energy	⊕ familiar patterns or arrangements (house, tree, table, etc.)	⊕ rubbing yourself with natural things

⊕ special settings	⊕ tasks of real importance and significance	⊕ floating on a small raft
⊕ meaningful missions		⊕ smelling things
		⊕ taking off some clothes

Take another look at our box on MAKING MAGIC in the section on Structuring for additional ideas on how to pull your learners in. Hookers and magic go together like bands and parades. (In fact, the original circus parades were designed as hookers to get the townspeople to literally follow the entertainers to the outskirts for the show.) Additional suggestions for immersers can be found in the box on IMMERSING TECHNIQUES located in the section on Immersing.

At this stage some notes of caution are needed. Since they are so powerful, hookers can also be hazardous. Here are some pitfalls to watch out for:

1. Some folks seem to get hooked on the hooker. I remember visiting one centre where the walls of the dining hall were literally covered with plaques signifying one unique thing or another, one odd record or another that different classes had achieved during their stay. Instead of using their hookers to motivate the kids in their educational programs, their program appeared to have become their hookers. Chances are good that what the kids would remember from their out-of-school experience would not be some particular instructional objective, but how they had set the record for some trivial event. After all, that is where a lot of their energy seemed to be focused.

2. Some leaders tend to make themselves the hooker (consciously or unconsciously). While this tactic may have some limited value with young teens, it's far better if you can make the program do that job itself.

3. Some activities may appeal to certain natural instincts or tendencies at certain ages (for example, live trapping and pond dipping may stimulate the

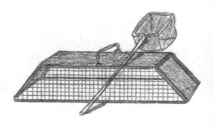

hunting instinct of young boys), but you will have to be very careful to keep such approaches from overwhelming your message (or actually demonstrating the very lack of respect that your message endeavors to overcome).

Organizers can also overwhelm your real objective. Since the organization-seeking mind of our species can be satisfied with most any pattern, the key is to keep it from locking in on nonsense. Our societies are already replete with examples of this unfortunate phenomena, so choose your organizers with care. And be sure you don't confuse the idea of an organizer with that of a theme. They are two different things. An organizer requires a lot more specificity. Besides, a themed approach to education is like learning how to play golf using only one club at a time. It's possible, but who is going to stick around long enough to learn the whole game?

Okay, if you have been champing at the bit to be creative in our design process, here's your big chance. Coming up with a hooker and an organizer, particularly ones that will work well together and enhance what we are after in earth education, requires lots of imagination. Just remember, people of all ages seek organizing patterns even when they are not aware they are doing so. (That may explain why some folks want to name everything. They are trying to organize things the hard way.) So in this respect you will be working with pre-motivated learners, but this means they will probably be more than happy to latch onto any organizer you provide for them. Consequently, you will need to make doubly sure that it actually fits in and supports your whole program.

"pROGRAM dESIGN: pROBLEMS aND SOLUTIONS"

There are two major problems with the design process we have outlined in this chapter. First, we obviously need to spell out more clearly all the pieces you can (and eventually must) include in an earth education program. In our example we used building a sense of wonder and a sense of place, but each of these perceptual tools represents only one part of those pieces under the Feelings that we refer to as joy and

reverence. So even in our example it is clear that you would have to do a lot more than you can cram into a half-day schedule if you were going to build a "joy at being in touch with the elements of life" and "a reverence for natural communities." And, of course, that was to be just the first step in building a whole earth education program. Once again, please don't forget, there is lots more to do.

Why is it that we so often let teachers and leaders come to our sites and centres with a group of young-sters like those in our example, spend a few hours with them, then go home thinking they have done the job required? Not so. In earth education we need to spell out for them what the big picture looks like and which piece of that picture we have been addressing. We need to say, "Here is the piece you were working on today, and here is what you need to do next if you are really serious about an educational response to our environmental crisis." It just won't do to let them leave thinking that their job is finished.

And why do all of our sites and centres try to do anything and everything themselves? Why can't we tell folks what pieces we are working on, and which ones we haven't tackled? "If you want this piece, come on out, we have been working on it for years, and we've got it really wired. However, if you want another piece instead, then you should call so and so down the road. We've worked this out in our regional association, and they're the ones working on that part." Doesn't this just make good sense?

Folks, let me tell you, building a good educational program takes a lot of time and sweat. There is just no way around that reality. So how did lots of centres get themselves in the position of trying to be everything to everyone? You know what I mean, someone phones them up with a request, and they say sure we can whip something up on that. And that's exactly what they do, often about 20 minutes before the group arrives. As I said in the beginning of the book, why can't we design and develop really good programs, then go out there and sell them to the teachers and leaders? Believe me, they will buy them. They want good stuff.

If you are interested in building a complete earth education program, or if you are interested in working on just a couple of the parts (perhaps getting together

YOUR HELP IS NEEDED...

Sadly, almost everyone out there appears to be supplementing these days, while practically no one seems to be programming. Sometimes I think every organization and agency and industry in existence is busy creating some sort of teacher pack of environmental materials. The field is literally awash with the stuff, and the teachers are being inundated with good intentions. For most of them, it is all they can do just to look through what is available, let alone figure out how to build programs with it, but you and I can. The nonformal learning folks that we represent could really do it. That's the educational challenge of our times. All the pieces necessary for getting started on some real programming are already out there — the places, the funds, the leaders, the learners — all we lack is the will. If you are in a nonformal educational setting, please consider building an earth education program and offering it to the teachers. They'll welcome the help (and we will too).

with other sites in your area to determine which ones), here is where you will need to start. We have broken the task down into 12 characteristics that must be developed for a healthy sense of relationship with the earth. They are the four items we examined for each one of the WHATS of earth education: the understanding, feeling, and processing.

THE BIG PICTURE

If we are serious about earth education, then we have to work at eventually developing all 12 of these characteristics in our learners. That is why I spent so many pages talking about each of them in the previous chapters. Obviously, it is going to take time to do this (we are still working on it in our own model programs), and you won't be able to do all of the job at each level of education. But the goal would be to see these characteristics instilled in learners of all ages coming out of all the various educational systems and settings in our societies.

What you can do personally depends upon your own setting and situation. The important thing is to get started. If you are working with pre-schoolers, for example, you may have to start by just looking at what you can do in those three broad areas represented by the head, the heart, and the hands. If you are working

"For a moment of night we have a glimpse of ourselves and of our world islanded in its stream of stars—pilgrims of mortality, voyaging between horizons across the eternal seas of space and time."

— Henry Beston

with upper elementary students you can begin by looking over the three programs we have already developed for those ages. If you are working with teenagers, you may be able to help them internalize the understandings a bit faster and move on to the feeling and processing points.

Whatever you do though, don't make the mistake of assuming that your learners have already internalized some of the basic understandings just because they are older or more verbal. Here's a marvelous classroom story that illustrates this error. A teacher had a geologist friend visit her class one afternoon and she suggested that he might like to ask the students some questions about the earth to sort of put them through their paces. He began with this query, "Class, if you dig a hole in the earth, as you get deeper will it get warmer or colder?" There were no responses. Everyone just sat there with these blank expressions on their faces. So he decided to try it again in a more descriptive way. "Just imagine that we could go out here on the school grounds and begin digging a hole. Well, as we dug down deeper and deeper into the earth, would it get cooler down there, or would it become warmer?" Again, nothing. Zip. Befuddlement.

Finally, the teacher intervened saying to her friend, "I don't think you are asking the question in quite the right way. Let me try. Class, in what state do we find the interior of the earth?" Ah, ha! Now they came alive, and in perfect unison responded: "The interior of the earth is in a state of *igneous fusion*."

Obviously, these students had learned some of the terms called for in the curriculum plan, but very little about their practical meaning. In our field I think you will find this true of many learners, regardless of their age. They may know some of the appropriate words, but they have never really assimilated or applied those words to their daily lives.

The second problem with our design process appears when you conscientiously work your way down through the seven steps only to discover there are no activities in the vehicle you have selected that will match up really well with your design criteria. Aaaauugghh!

To make matters even worse, we are going back now to revise some of our own activities in order to make sure we are doing the job ourselves. Fortunately, our activities for the ecological understandings were always developed with specific concepts in mind, but on the feelings side we don't have that much specificity worked out yet. So in some areas you are not going to find a lot of depth in the choices we can provide at the moment.

In fact, I am afraid at some points in the activity selection phase you will have to decide whether to sharpen or revamp an activity from some of the supplemental materials available (after all, they invite you to do anything you want with their stuff), or work on coming up with a new one yourself. (Of course, in the latter case, we are hoping the institute can provide a way for you to share your results with others.)

You have probably guessed it by now. I don't like the word "adapt" very well either. In our field, it often means either an activity has been skewed around to such a point that it can no longer really accomplish what it was designed to do, or just enough changes have been made to justify publishing it in somebody else's collection.

No, that doesn't mean we are insensitive to the need to make some adjustments in our activities in order to meet the requirements of different settings and situations. In fact, that's the term we prefer to use: adjust instead of adapt. The difference is in the degree. It just seems to us that at some point an adapted activity often becomes something other than what it started out to be. Nor does it mean we are unwilling to share. We just hate to see good activities torn from their context and used without all of the components that are required to make them work really well.

Obviously, it would be helpful if you could find some of the activities required for your program among all those supplemental collections that have already been produced. Just remember, though, that you should figure out what you want to accomplish first, then go looking for activities afterwards. As I noted before, in our design process that step is very different from most other programming approaches. Lots of leaders practically reverse the sequence. They start out looking at the activities instead of ending up there.

So how can you tell if an activity that is already available, or one that you created yourself, will make a good earth education activity? Here are some questions to ask yourself at various stages of the learning experience.

WHAT MAKES UP A GOOD EARTH EDUCATION ACTIVITY?

QUESTIONS TO ASK AT THE START

Does this activity:

☑

☐ Fit smoothly within the "big picture" of earth education principles?

☐ Match up with the criteria of an existing earth education vehicle?

☐ Hook the learners and pull them in?

☐ Deal with fundamental ecological processes?

☐ Establish a friendly, non-threatening atmosphere (even if it is a challenge)?

☐ Set the learners up for what they are going to do?

☐ Explain or reinforce why and how its outcome is important?

QUESTIONS TO ASK DURING THE ACTION

Does this activity:

☐ Include some firsthand contact with natural systems and communities?

☐ Contain a key statement that summarizes the "essence" of the understanding or feeling?

- ☐ Stand on its own without a lot of "discussion" to get its point across?

- ☐ Follow an organized, carefully-crafted sequence of steps?

- ☐ Contain built-in ways of reinforcing its outcomes?

- ☐ Involve all the learners in ways that are appropriate to their age and ability?

- ☐ Get the concept into the concrete?

- ☐ Emphasize magic and meaning (M & M's) instead of names and numbers (N & N's)?

- ☐ Achieve suitable physical (hands-on) and mental (minds-on) engagement?

- ☐ Include "focal" points whenever the leader is talking for more than a minute?

- ☐ Use terms and analogies common to the learner's experience?

- ☐ Use questions to facilitate the flow of the action (rather than quizzing the kids)?

- ☐ Focus on specific outcomes that the learners can identify beforehand and relate to afterwards?

QUESTIONS TO ASK AFTERWARDS

Does this activity:

- ☐ Result in an outcome that will be useful to the learners later in life?

- ☐ Require the learners to do something with what they have done?

- ☐ Fit carefully into the overall sequence of the program?

- ☐ Reward the learners for successfully completing it?

- ☐ Accomplish what it set out to do?

Naturally, this checklist will limit the number of activities that you may end up with after culling through all the supplemental collections available, but I am afraid there is no easier way to determine what will make a good learning experience for an earth education program.

Typically, we spend several years developing a program ourselves, and we toss out lots of activities and ideas in the process. For instance, during the final stages of the development of our Sunship Earth program, the staff decided to include a night hike as a special late evening event. To their surprise, I objected. It seemed to me that the traditional night hike didn't have enough of either the magic or the focused learning that we sought. Oh, some adventure was there since it took place at an unusual hour, and some learning was there based on what the counselors pointed out and talked about along the way, but I wanted something more. I wanted a magical experience that pulled the kids in right from the start.

We began by trying to boil down what we really aimed to accomplish in a night-time activity, then proceeded to look for ways to get those points across. It took us four years to come up with an alternative. I remember our first attempt almost too well. It was called, "The Little Owl who was Afraid of the Dark." But after all the hours we put into it, all the effort and sweat, it just didn't have any magic, and in the end we had to dump it. (We should have known we were in trouble I suspect just with the name.)

So we started over. The second activity we came up with was known as "The Deer Stalker." We thought we were on the right track with that one. It seemed to have elements of both the magic and the adventure that we required. Anyway, we sent it off to be piloted, but the kids merely shrugged their shoulders without much enthusiasm.

Finally, on our third try, we came up with the idea that there were "nightwatchers" out there who were in charge of the changing of the guard as the earth turned away from the sun. We explained to the kids that one of these people would be coming to their cabin after dinner sometime to introduce them to this special time of day. You can imagine the kids' reaction when someone largely hidden by a black hooded cloak and carrying an old candle lantern suddenly appeared one evening scratching on their door. . . .

The point is, we were willing to discard those activities that didn't have the necessary magic or adventure or educational focus. After all our work in developing a new activity, we would stand back and ask ourselves if it was really doing the job, and doing it in the way we wanted it done. If not, we would toss it out. (And we are still tossing.) Frankly, I always suspected that that was a major element in what made us so different from many of the other groups that were creating activities in this field. We were more willing to dump the marginal ones.

Are there any restrictions on using some of the other activities we have already developed in the institute? No, of course not. When you buy them, they are yours. However, there are some legal restrictions on duplicating our copyrighted materials or in using our trademarked titles and logos. (We maintain that only those who have purchased a program package from us and are offering a complete program are entitled to use the latter.) Over the years, there has been a lot of misunderstanding about our position on all this. Unfortunately, we became a bit paranoid early on about the situation ourselves and probably sent out mixed messages as a result. So I know we are partly to blame for the confusion.

At any rate, as we have explained in our seven step design process, there is nothing wrong with taking some of the activities from our various sources and using them to build your own earth education program. We only ask that leaders use them in their entirety, at an appropriate site and with the necessary props. However, there are two major exceptions to this. . . .

First, the Concept Path activities in the Sunship Earth program were designed using the I-A-A learning model. Once again, the learners take something in (Informing), do something with it (Assimilating), then use it (Applying). At each station on a Concept Path the learners begin by reading the appropriate input in their "Passports" or learning booklets for that particular ecological concept. Next, they participate in a focused activity that will help them bring the concept into the concrete and internalize its meaning. Afterwards, they must use their new understanding by finding and recording an example of that concept operating in their immediate natural setting.

Obviously, if you conduct the activity without the passport input (the informing level) and the task afterwards (the applying level), or in a setting where it is difficult to feel that those ecological processes are larger than we are, then the learners will miss out on much of the power of the learning experience. It just won't work very well to pull the activities out of their context and plop them down somewhere else by themselves. (That's why you will need to obtain the copyright permission so you can print up your own learning booklets.)

By the same token, we designed the Earthwalk activities to be used in conjunction with one another. We don't believe that just one or two of those activities, inserted in another experience, can begin to build the sense of wonder and relationship we had in mind. As a result, we ask people to put together complete walks (at least 4-6 activities) and set aside the appropriate time (about 45-75 minutes) to guarantee that this will be an important experience.

In addition, our encouragement for building your own program does not mean we think it's okay to use our hookers and organizers to patch together a bunch of other outdoor activities, or to call a bunch of other activities by the name of one of our model programs. To give you an idea of what can happen, here is just one of the many situations we have had to face in the past few years.

An outdoor centre in the midwest set up their own version of Sunship Earth by organizing their existing activities (plus adaptations of some of ours) around our EC-DC-IC-A formula. They started the kids off with a videotaped edition of "Cosmic Zoom," then worked their way through a traditional potpourri of outdoor activities (including everything from studying stream flow and timber cruising to stops at the weather station and the trapper's cabin). They had none of the key elements that the Sunship Earth program is based upon — no Concept Paths, no Passports, no opening and closing ceremonies, no daily Magic Spots, no Interpretive Encounters, no Sunship Meeting, no Passengers Guide, etc.

After spending quite a bit of time working with them about how they had dismembered our carefully crafted program, we finally had to insist that they change their name. As you might suspect, they were not pleased. They just did not seem to understand our fear that other people would see this vitiated version and think it was the real thing. In our eyes, students, teachers, parents, interns, visitors, etc. would likely come to their program and leave thinking they knew what Sunship Earth was all about, and naturally, we believed that was unfair to both us and to those centres who were actually doing the program. (Of course, you can't really "win" on these things, for they simply started calling it "Earthship Journey" instead.)

I hope you can appreciate our dilemma: we want to maintain the integrity of earth education programs and activities, *and* we want to help people build earth education programs that will fit the needs of their own settings and situations. We have never said it would be easy though. We have spent years working on our own model programs, and we suspect that it will take just as long, if not longer, for anyone else to do it. Unlike other folks in this field, we are just not interested in providing some activities to spice up an occasional outing. We want to spend our efforts helping those who are serious about the mission of earth education. In short, we hope you will go for the real thing and call it that if you do.

"program leadership: keeping on target"

In the opening chapter, as part of our explanation of how environmental education went astray, we included that old saying, "If you aim at nothing, you'll hit it every time." By now you should have a good picture of our earth education target, but as a program leader how can you hit that bulls-eye every time? Well, maybe hitting the bulls-eye every time is too much for any of us to expect, but here are three things you can do to come closer to that goal. First, put a lot of effort into preparing your staff beforehand. Second, stay on top of any changes to be made enroute. Third, when they are finished, reward your leaders for giving it their best shot. In the following pages, we will look briefly at some helpful hints for completing each of these tasks.

First, you need to spend as much time in training your staff as they spend working with the learners. In addition

to having them study this book, and taking them through the appropriate activities, spend some time arranging a special natural experience for them (Remember: "If we care, they'll care"). Also, ask them to roleplay various problems they may encounter in their activities (like the learner who doesn't want to get dirty or the participant who says he has done this before). After that, it's practice, practice, practice.

One of the questions we get asked a lot by prospective leaders is what to do with disruptive learners. To be honest our programs have such powerful hookers that this is seldom a problem. (In fact, if it were, that would be a sure sign that there was probably something wrong with the activity.) However, when our learners do go astray, we usually just ask them to sit and wait until they are ready to join in again. The ensuing action almost always pulls the occasional problem participant right back in within a minute or two.

Naturally, we could write a whole book on staff training alone, so to keep this one manageable, I will just include some special roles that we believe you will find extremely helpful for organizing your leaders. Filling the following positions will go a long way in keeping your program on target:

Keeper of the Magic — *someone who will keep an eye open for opportunities to add a bit of magic to the usual routines, yet make sure no one gets carried away and adds too much.*

Prop Manager — *this person will prove invaluable for scrounging up needed materials, maintaining inventories, making small improvements, organizing work sessions, and handling repairs.*

Details Director — *the anti-entropy specialist makes sure that activities have not succumbed to unnecessary modifications. Keeps everyone on target and spot checks for problems. Cautions colleagues about sidetracking influences, but helps prevent staff burnout by putting energy into those important supportive details that easily get overlooked.*

Keepers of the Secrets — *parents and visitors can take on the important task (between activities) of listening to the participants explain the key points to clarify and reinforce their learnings (just make sure they understand that this is a limited role!).*

"Genius is childhood recaptured."

— Baudelaire

Second, you need to carefully think through and monitor any adjustments made in the activities themselves. We have put together some case histories, real examples of leaders altering earth education activities and programs, to help explain this task. Some of the cases represent good ideas which enhance the point of the activity involved; others, however, describe negative examples, alterations which distract from the learning. Read them through, and based on what you know now about earth education, see if you can figure out how we would reply. (Our responses are located at the end of the list.)

A. A small group sets out on a Concept Path at a Sunship Study Station. The leader has them begin chanting loudly "EC-DC-IC-A, EC-DC-IC-A!" (the organizer for Sunship Earth).

B. A leader decides to eliminate the strings from Micro Parks in favor of using some sticks lying around on the ground. "Just grab a handful of sticks and lay them around your park for the border." The leader explains later that this makes it easier to prepare for the activity and works just as well.

C. In addition to reading the Passport entry before the Concept Path activity, a leader decides to have the kids read it again at the end of the activity, after the name of the concept is stamped on the "example" page.

D. The leader begins a concept activity with the question, "Does anybody know what 'cycling' means?"

E. The leader ends the concept activity with the question, "What does 'cycling' mean?"

F. A leader is anxious to create excitement for his new group of learners. He wants them to see him as a fun and unusual character. So he meets the group to set the stage for "Welcome Aboard" (a serious ceremony) dressed in a long black cape with a wizard's hat.

G. An enthusiastic Sunship Earth leader spends considerable time building the leaf for "Food Factory." Just about every aspect is improved upon, including the use of a 20 foot long 3 inch drain pipe with 3 foot extensions of flexible tubing on each end (for the secret speaking tube of "leaf control"). The result works perfectly, but the "new" leaf has a lot of visual impact on the area.

H. To get the learners "into" the "Border Dispute" activity a bit more, the leader builds up the competitive spirit by having one team denounce and taunt the other.

I. Recreation activities ("Time Out") are promoted all day long as motivation to keep the participants' attention during the other portions of a program.

J. A visitor at the Sunship Study Station dresses up in an official looking costume, complete with Sunship badge and sash, plus a hat which says, "Sunship Study Station Border Patrol." She stops the groups between stations and says: "Excuse me — official Border Patrol here. I'm going to have to briefly check your Passports to make sure you are all bona fide passengers on Sunship Earth. Carrie, can I see yours? Uh-huh. Well, I have to make sure this passport is yours. What do you mean here by 'Community?'"

K. A leader is interested in making the job of the high school counselors easier. He streamlines the Cue Cards for the activities so the tasks are just listed on one card.

L. The leader leaves notes in little jars and bottles for the participants to find between activities. The notes illustrate the concept they just learned by applying the idea to something nearby.

M. It rains hard all morning. The learners participate in outside activities for the first half hour, then go inside and play games.

N. A visiting school has a tradition it wants to continue. They always have each youngster bring a small bag of soil from his or her area, then they ceremoniously plant a tree seedling with the accumulated soil just before the students board the bus to return. Now they want to do this after the closing ceremony in the Sunship Earth program.

a. Although the leader should be recognized for her positive energy, the idea itself is not so good. EC-DC-IC-A has been introduced in "Welcome Aboard" as a special formula for the story of life on earth. Making it into a sing-song chant will probably detract from the seriousness of the mission. Besides, the leader should be accompanying the learners, not directing them, on their adventure.

b. The activity probably still works, but some problems are likely to result from the change. First, the stick method is a little abstract for some of the younger learners to grasp, whereas with the string in hand the border for the "Micro Park" is ready to go and will likely be used. (And since it is to be laid out in a circle, it establishes a manageable size as well.) But a greater concern is where this modification may lead. Next, the station markers may be dropped from the list of props because "they can just use sticks for that too." Then the peanuts may be eliminated in favor of "informal sharing," and finally the leader may report, "You know, that Micro Park activity doesn't really work very well." Actually, an enthusiastic leader could probably pull off this activity without the props, but it would take a lot more personal energy each time to do it. In addition to helping with the motivation for the learner's closeup view of the natural world, the props make the job of the leader a bit easier too.

c. This one is a good idea. It enable the learners to further clarify the concept. However, time constraints may prevent the leader from doing this after each activity.

d. This leader is falling into the "Twenty Questions" trap. Of course, most learners won't know the answer — they haven't developed that "filing folder" yet — for that job belongs to the activity, and the leader should let the activity do its job.

e. While this is not such a good practice, it is obviously better than the previous example. Here the activity has come to a close and the question is used to help the students reinforce the point. Still, it is posed from a "teacher-tester" angle instead of a "learner-helper" one. A better way would be to wait until the group meets up with the Station Leader or the classroom teacher and say, "Hey, you guys, let's tell Mrs. Hobbs what we learned about cycling."

f. Not so good. It is the beginning of the program and the learners aren't quite ready for it. The job of the leader was to set a serious and concerned tone for the ceremony. However, before he was able to do this, the kids started pointing and laughing at him. Lighter and more active parts of a program may lend themselves to costuming and "hamming it up," but not so for the serious moments.

g. Nice going. Often the best way to channel excess leader energy is to have them improve upon what you know works well instead of experimenting with the unknowns. Besides, visual impact in this case actually enhances the appeal of the activity and adds to the suspense. Now, turn her loose on the "Best Deal on Earth. . . ."

h. Be careful here. The leader should be ready to step in if the competition gets out of hand. A few trolls and elves have been known to end up with bloody noses when the competition was overdone. If it gets to be a bit much, the wizard can intercede with the line, "Okay, as the official wizard here, I have a new law. No arguing, yelling or fussing on scientific missions. Any Elf or Troll that disobeys will immediately be turned back into a fifth grader!"

i. Not a good idea. Students traditionally view recreation as fun and learning as boring. If they are fired up about the recreation ("Time Out") activities to come, they may miss out on the fun of what they are doing in the present. After all, earth education activities are designed to be fun too and the learners should have a chance to see them as such.

j. Fine. This is an acceptable way to use questioning. The leader has "become someone else" and chosen a fun way to ask the questions. (If it is even slightly threatening though, it should be softened with a bit of humor.) The question helps the learner repeat and reinforce the idea of community. This is also a fun role for teachers and volunteers to play. Be careful though, too much magic can work against the goal of reinforcement (such as wearing a mask so the learners wonder more about who it was instead of processing the message).

k. Watch out! Counselors may skip vital parts of the activity. If you streamline what they have to say or do, they may eliminate it entirely. The key to solving the original problem may be to allow for more *preparation* and *practice*, or to recruit more experienced counselors.

l. Good reinforcement. The participants have a fun way to review what they have learned without being tested or quizzed by the leader. But why not ask the learners to leave some notes too?

m. Not good enough. The leader was not prepared well enough. Alternative schedules and activities that support the goals of the outside activities should have been ready. Inclement weather may occasionally pose a challenge, but "winging it" is a weak way to meet the challenge.

n. In this case tradition gets in the way. Two ceremonies in a row will only minimize both of them. We should not "water down" our programs just to make them more appealing to more groups, but make a case instead for what we are trying to accomplish and convince people that these programs have great value in their own right. If this school must include their soil ceremony, then give them a time slot early in the morning or the day before at "Quiet Time." We must consistently scrutinize everything done and constantly ask, "What's the point?"

Before we go on to our next point, let's sum up how the leaders in these cases managed to get off-target:

⊕ falling back on "traditional" techniques: questioning, lecturing, discussing, identifying, etc.

⊕ streamlining by skipping parts or eliminating props

⊕ getting off on the magic

⊕ holding on to a favorite idea and trying to fit it in even though it was inconsistent with earth education

Third, you need to reward your staff for remaining at the top of their form as leaders. Earth education activities include built-in elements of adventure and action, but the leader must play the all-important catalyst for everything to work well. Frankly, it is easy for newcomers to lose sight of our objectives and find themselves off-target — it even happens to us old timers occasionally. That's why it is important to ask them to stop, back up, and take a look again at what they are doing.

Here is a checklist that you should have your leaders run through from time to time. (Consider giving each item points and asking your leaders to rate themselves, but be sure to reward them for improvements.)

"We have two ears, but only one mouth in order that we may listen more and talk less."

— Zeno

eARTH eDUCATOR'S pERSONAL lEADERSHIP iNVENTORY

☑

☐ I put ample time into preparing every facet of an experience, including site setup, props, leader's role, timing, etc.

☐ I demonstrate a sense of wonder and awe for the natural world.

☐ I motivate learners with magic and adventure, not by using threats or withholding rewards.

- ☐ I put some ENERGY into my role as the leader, but refrain from making myself the star.

- ☐ I always "set the stage" for what's coming and summarize the action afterwards.

- ☐ I make sure that every learner is given a participatory role in every activity.

- ☐ I constantly make an effort to personally reinforce the learner's new understandings and discoveries.

- ☐ I include focal points whenever I am talking.

- ☐ I pay careful attention to the details of each activity.

- ☐ I use verbiage sparingly during an activity.

- ☐ I emphasize questions that facilitate action instead of quizzing the learners.

- ☐ I avoid naming and labeling things.

- ☐ I practice "sharing and doing" rather than "showing and telling."

- ☐ I make sure that the activity does the teaching instead of me.

- ☐ I find lots of ways to help the learners relate things to their own lives.

- ☐ I enjoy the learning activities as much as the learners do.

- ☐ I model good environmental behaviors.

"He who has begun his task has half done it."

— Horace

Okay. That's enough. Think of all the leadership guidelines that we have touched upon both here and in chapter five as a pouch of earth education tools that you will carry around with you. Like any craft you will find that your skill in using them will improve over time. The important thing now is to familiarize yourself with them well enough that they will be readily available for you when you need them.

"THE
EARTH EDUCATION
PATH"

This chapter is probably the most important part of this book. Actually, I feel a little like Robert Pirsig who said he finished <u>Zen and the Art of Motorcycle Maintenance</u>, then threw everything away except the last chapter and that became the opening of his final work. So please don't get bogged down in all the details in the beginning parts of this book.

The critical thing now is to get started. Get out there and set up an earth education program somewhere yourself. And don't wait on us. Sure, we would like to hear about it someday (we might even be able to help), but that doesn't really matter. What matters is just to get on with it. The earth can't wait. (Just be sure it is a genuine earth education program, not an environmental or outdoor education one.)

I closed the introduction to the book with a request that you consider joining us on the path of earth education. Now that you know where we are going, if you would like to undertake that journey with us, we have put together a visual map to remind you of the tasks along the way.

First and foremost, think of this as an adventure. You are setting off to discover a more harmonious and joyous future (for yourself, for those you work with, and for the other life of the earth). At the end of the journey you will have crafted a lifestyle that incorporates thinking globally, acting locally, being personally.

Second, we have posted symbols for the head, the heart, and the hands along the way to remind you that an earth education program must integrate all three of these components (the understanding, the feeling, the processing) into one continuous flowing experience. And you must do the same in your own life.

Third, special signs for the hooker, the organizer, and the immerser will focus your attention on these important tools, for without them you and your learners will probably lose your way. (And they should help

"Our grand business is not to see what lies dimly in the distance, but to do what lies clearly at hand."

— Thomas Carlyle

A WHOLE BODY APPROACH...

Please remember that the head, the heart and the hands must work smoothly together just like they do in your body. It will not do to tack the hands on as an afterthought or assume that the head and the heart are in synch just because you have included activities dealing with both. As a leader your most important task will be to make sure that these components become synergistic so that the whole person responds to earth education in a life-changing, life-sustaining way.

Frankly, this takes a lot of time, a good eye, and an intuitive feeling for the task. Don't be reluctant to ask for some help. We have a lot of program specialists in the institute who would be willing to review your plans and provide some feedback.

THE EARTH EDUCATION

HOOKER

ORGANIZER

LEARNING to

PATH

thinking globally
acting locally
being personally

IMMERSER

IVE lIGHTLY...

J. Munr.

you avoid the pitfalls and false trails carved out by those who have been led astray by supplementalist and cornucopian thinking.)

Fourth, if you look closely, you will see our all-important secret ingredient hidden in the pathway itself, where it can provide an underlying reminder of what makes our work so special, and why you are so special in that work.

Good journeying.

"tHE iNSTITUTE fOR eARTH eDUCATION"

In 1974, a handful of folks interested in helping people build a new sense of relationship with the earth formed an organization that, much to their surprise, would end up having a significant impact upon the nature of outdoor learning in our societies. There were seven charter Associates: Jim Wells, Dick Bozung, Harry Hoogesteger, Oliver Gillespie, Donn Edwards, Brenda Slickman, and myself.

I was on the faculty of George Williams College at the time and decided to put together a series of Acclimatization (ACC) Workshops at our Lake Geneva Campus in southern Wisconsin. Needing some help, I invited Jim Wells and Pat Walkup, who had worked with me at the summer camp where ACC had its origins, to come to Lake Geneva to assist with that effort.

Large enthusiastic groups, made up of leaders and teachers from across the country, turned up for those initial sessions and sparked the first interest in forming some sort of association. In addition, *National Geographic* magazine carried a story on our work about the same time, and Acclimatizing was released shortly thereafter. Week by week it became more and more obvious that a formal base of operations was needed to continue developing and sharing the ACC idea. Finally, using some of the royalties from the first books, we set up shop in the ACC Workshops Office, and the rest, as they say, is history. Over the years, those seven original Associates have increased to a working core of 150 volunteer representatives, plus hundreds of other kindred spirits in major branches in the United States, Canada, Britain, France, Australia and New Zealand.

Although the intent from the beginning was to build an organizational home for a particular school of thought about nature education, as opposed to becoming merely another umbrella group, the institute has now outgrown other professional associations in its field to become the world's largest group of educators devoted to helping people live more lightly on the earth.

Our model programs for 10-12 year olds (Sunship Earth, Earth Caretakers, Earthkeepers) already reach tens of thousands of learners annually, while the development of SUNSHIP III (ages 13-15) will be completed this year, and the piloting for Lost Treasures (ages 8-9) will begin.

In short, it is probably no exaggeration to say that our work has literally changed the face of nature education. Twenty years ago, when <u>Acclimatization</u> was first released, many of the things that are now taken for granted in this field simply were not included in most educational offerings. Victorian nature study with its emphasis upon identification and collection, along with a smattering of nature crafts, survival techniques, and Sputnik-induced interest in science experiments (plus the usual capturing and dissecting urges), dominated the outdoor learning scene. The Acclimatization program was widely recognized as a major new approach, the first real breath of fresh air in the field in a long time.

Here are a few of the institute's pioneering thrusts in nature education (some of which may still be ahead of their time):

⊕ *developing "acclimatization" immersing techniques*

⊕ *building carefully-crafted, structured programs*

⊕ *emphasizing primary ecological concepts*

⊕ *using solitude experiences as regular, integral parts of a program*

⊕ *avoiding labeling and quizzing strategies*

⊕ *utilizing extensive props and "magic" in total educational experiences*

⊕ *creating educational "hookers" and "organizers"*

Fortunately, some members of the institute volunteer each year to serve as representatives and facilitators for earth education in their areas. They are the working core of the organization. Everything that is done in the institute is done by one of these Associates and it is this wonderful group of selfless people that has made possible all of the developments shown here.

Never doubt that a small group of thoughtful, committed citizens can change the world. Indeed, it's the only thing that ever has.

— Margaret Mead

⊕ *designing programs that must be completed back at home and school*

⊕ *instituting specific tasks and pledges for changing environmental habits*

⊕ *publishing and distributing complete program packages*

⊕ *establishing an international series of training workshops and conferences*

⊕ *accrediting quality programs and centers*

⊕ *developing a worldwide network of program leaders and educators*

It's hard to believe, but prior to the institute almost none of these existed in any substantial, organized sense in the general area of outdoor learning. And what's really amazing is that we did all this without the usual funding. There were no government contracts, no foundation grants, no industry sponsorships — just the personal energies and contributions of our volunteer staff Associates who refused to give up their dream of a new way to educate people about the natural world. And after all these years, their generosity and commitment still sustains us.

Today, The Institute for Earth Education is a nonprofit, grass roots association made up of an international network of individuals and member organizations. Our primary work is to support the design, development and dissemination of specific educational programs that change people's view of their home, the planet earth, and the way they interact with it. Our primary funding comes from the sale of our materials, fees from our workshops, dues from our members, and small contributions from our annual appeal.

How do we differ from other environmental groups? No other organization in this field exists solely to develop and disseminate quality educational experiences. True, lots of groups have created collections of supplemental materials, but as you have seen, we are supporting the creation of complete programs. Lots of groups serve as umbrella organizations for those with a wide range of outdoor or environmental interests and intentions, but we are chartered primarily for those interested in the educational process. Lots of groups are working on the present problems, but we are preparing people to deal with the future ones. We are convinced that focused educational programs must serve as the seedbed in the years ahead for the personal change and advocacy necessary to preserve this fascinating planet we share.

We believe:

⊕ Earth education should be a separate and distinct part of every school curriculum, youth program and adult organization.

⊕ Understanding basic ecological concepts (and their meaning in our daily lives) is too important to leave to a chance lesson or activity or talk.

⊕ Heightened feelings for the natural world combined with increased understandings about its systems and communities form the foundation for positive environmental action.

⊕ Learning experiences in our field should include more "M & M's" (magic and meaning) and less "N & N's" (names and numbers) and take place primarily in natural settings.

⊕ All environmental programs should require their participants to begin making personal improvements in their own environmental habits, while insuring that their leaders and sites serve as models themselves.

⊕ Earth education is a serious task, but getting to know the earth should be a lifelong adventure full of wonder and joy.

If you share such beliefs, please contact us for a packet of membership information. We think you will find many kindred spirits in our growing earth education family.

THE TIDE HAS TURNED...

An international survey conducted last year for the United Nations found that a majority of people around the world believed government should do more to protect them from pollution. Obviously, there has been an amazing resurgence of interest in our environmental condition. On the environmental action scene there are lots of positive developments as the wave of concern has begun to build once again (with new groups, new publications and new projects popping up everywhere along its surface). So there is much cause for rejoicing.

However, on the educational side I fear that many teachers and leaders may get caught in the process in a strong "infusion" undertow and be sucked down unaware once again into the muck of inanity and inaction. Please warn them: don't infuse, integrate. Encourage them to build genuine learning programs and work them into their school curriculum or organizational calendar. Ask for change. If we are to ride this new wave of concern as far as it will be necessary this time to make a lasting difference, we must not be satisfied with a few externalizing discussions and short-term projects in our classrooms. The earth needs your best effort on this, for it will be the educators who will really make or break the long term environmental outcomes of this new opportunity.

THE EARTH EDUCATION TREE

Our structural symbol is based upon a magnificent tree that lives in northern New Zealand. It is one of the largest living things on earth, and its unusual appearance to a North American or European eye reminds us of the international scope of the institute.

Now that we have introduced you to our path perhaps you will be interested in establishing a local branch of our international tree. It only takes two or three people.

Of course, the first step is to become a member of The Institute for Earth Education and get more "grounded" in what we are all about. Next, the coordinator of a local branch must also be a part of our worldwide volunteer staff; so someone will need to consider becoming an Associate representative of the institute. This is an important role that allows us to keep in touch with what's happening and provide the appropriate support and guidance.

Local branches serve as catalysts for earth education in an area. Instead of meetings they have working sessions in which they focus on building programs, experiencing the natural world, and living more lightly on the earth. If you would like more information, please contact us for a Local Branch packet.

"The first day or so we all pointed to our countries. The third or fourth day we were pointing to our continents. By the fifth day we were aware of only one earth."

— Sultan Bin Salman al-Saud
The Home Planet

"To live content with small means, to seek elegance rather than luxury and refinement rather than fashion, to be worthy not respectable and wealthy not rich."

— William Ellery Channing

ePILOGUE

Welcome home. It should be obvious by now, we believe that many of those people today who call themselves environmental educators are actually doing other things. The term environmental education has been so misused and abused that we have given up on it. Perhaps someday that area of the field will sort itself out and begin rectifying its current image as merely an umbrella for outdoor interests. We wish them well. We even hope our work might be of some assistance in their efforts.

Meanwhile, we want to get on with the urgent task of earth education. It is an alternative designed specifically for those who are serious about an educational response to our environmental crisis and are willing to make the personal sacrifices necessary to see genuine, broad-based educational programs implemented for that purpose. The goal of helping people develop a better sense of relationship with the earth, while learning to live more lightly upon it, is too important to let it get lost out there amidst a plethora of unrelated groups and committees and reports. We are making a home for it. Won't you join us?

"A vision without a task is but a dream, a task without a vision is drudgery, a vision with a task is the hope of the world."

— Church Inscription Sussex, England, 1730

LEARNING TO LIVE LIGHTLY

eARTH tAX

Our tax stamp denotes that we have used a portion of the price of each book to offset the additional costs often required by our present economic systems to engage in more environmentally sound practices (such as printing on recycled paper).

iNDEX

a

b

barriers: 58, 95, 258; attitudinal, 54; personal, 263; mechanical, 55, 57
bathtub, 65
baubiologie, 133
behavior: change, 3-6, 8, 19, 21-23, 25, 31-33, 37, 42, 88, 97, 101, 129, 131, 158-159, 165, 249, 252, 267-268, 270, 279, 312-313; modeling, 87, 169; patterns, 21, 114-116, 146-147, 279
Belgrade Charter, 11
big picture, 3, 8, 17-18, 20, 32, 105, 132, 276, 278, 289-290, 293
billboard messages, 276
billion, 109, 111, 114, 117
biology, 125, 130, 133, 249
birding, 6, 129
blindfolds, 10, 84, 193-195, 286
bog, 53, 179-181
Bookchin (Murray), 99, 150
bookshelf (earth educators), 148-155
Border Patrol, 301
boxes: in suburban living, 70-71, 95; in Earthkeepers, 284
branches (earth education), 251, 310, 315
breathing, 111
Brown (Lester), 92, 151
Bruner (Jerome), 280
Bureau of Land Management, 144

C

camp, 52-53, 59, 61, 65, 74, 76, 134, 193, 236, 241, 310
candy, 249-250
Capra (Fritjof), 98, 149
caring (in magic), 199-200
Carson (Rachel), 123, 125
cemetery study, 25
ceremony, 67, 256, 300-304
ceremony of the marsh, 59-60
challenge, 32-33, 67, 158, 202, 208, 213-214, 286, 293
change: behavioral, 3-6, 8, 19, 21-23, 25, 31-33, 37, 42, 88, 97, 101, 129, 131, 158-159, 165, 249, 252, 267-268, 270, 279, 312, 313; ecological, 98-99, 105, 115-119, 164, 214, 232, 234, 269, 277-278; educational, 29, 31, 178, 188, 196-197, 292, 298, 302, 311; perceptual, 312
characteristics (program), 269-270
chickens, 133-137
Chief Seattle, 256
childlike, 198, 228
climax forest, 193-194
Closing Circle (The), 90
clothing, xi, 67, 133, 137, 143, 199, 218, 220, 223, 234, 259
Cobb (Edith), 228
collecting, 32, 94
colony, x, 70

d

discuss (colon), 192
discussion: 9, 25, 138-139, 207, 250, 252, 268, 294, 305;
    environmental problems, 19, 24
disguise-removing, 133-137, 213
Disguise-Removing Techniques, 137, 262
disguises: educational 36-43, 144; perceptual 133-137, 139
Disney, 126, 258
Disney World, 78-80, 200
dissecting, 311
distilling essence, 230-231, 274, 276, 293
diversity: as disguise, 37-38; ecological, 98, 113, 165, 232, 278
doing: 62, 66, 68, 101, 182, 184, 186-187, 204-209, 235, 280;
    techniques, 206
dolphin, 191
Dubos (René), 19, 159

e

E.M., 257, 284
earth bashers, 268
Earthborn, 226, 283
Earthbound, 283
Earth Caretakers, 44, 200, 255, 257, 283, 311
earth champions, 87, 97, 100-101, 140, 268-269
earth education: activities, 293-294; branches, 251, 310, 315;
    characteristics, 269-270; path, vii, ix, 307-309, 315;
    principles, 87-88; programs, 283; pyramid, 86;
    tree, 314-315; vehicles, 262-264; vs. environmental
    education, 47, 86, 252
earth educators: 96, 67, 100, 157, 228, 244; bookshelf, 148-155
Earth Journeys, 58, 262
Earthkeepers, 244, 256
Earthkeepers, 44, 147, 255, 257, 278-279, 283-284, 311
Earthkeepers Training Centre, 256
Earthlings, 283
Earthwalks, 58, 84, 142, 259, 262, 264, 297
Earthways, 283
Earth Speaks (The), 96, 142, 255, 277
EC-DC-IC-A, 26, 257, 278, 297, 300, 302
ecocentric, 130
ecological: communities, 105, 113-115, 125-127, 174, 176, 232,
    254, 259, 270, 278, 301, 304; concepts, 6, 17-18, 23, 25,
    33, 47, 62, 64, 93, 102, 132, 139, 178, 213-214, 249, 254,
    262-264, 269-270, 272, 274-275, 278, 283, 292, 296, 311,
    313; feeling, 33, 93-95, 98, 140, 233, 252, 281-282;
    processes and systems, 3, 5, 9, 14, 20, 24-25, 87, 105, 127,
    134-135, 137, 213, 221, 293, 297; vs. sociological, 10, 12
Ecology of Imagination in Childhood (The), 228
economics, 97, 149-150, 154, 165
Einstein (Albert), 48, 101
electric power companies, 34

f

g

h

habits: 10, 20-21, 101, 146-147, 158, 164, 203, 215-217, 223-224,
  263, 267, 283, 312-313; tasks, 146, 263
Hardin (Garrett), 110
harmony, 3, 83, 99, 124, 141, 144-146, 148, 153, 232, 236
Harmony Houses, 157
head, heart and hands, 102, 164, 291, 307
healing, 95-96
hemlock, 174-175
hidden messages, 8, 34-36, 252
high school, 20, 67, 132, 233, 257, 301
hobbits, 260
hobbyists, 28
hookers: 253, 255-257, 260, 269, 275, 282, 288, 297, 299,
  307, 311; techniques, 284-287
Hucksters in the Classroom, 34
hugging trees, 128, 233, 258, 286
human (role), vii, 87-90, 125, 127, 129
hunting, 113, 233, 288
Huxley (Aldous), 69, 158, 228

i

I-A-A (learning model), 279-280, 296
identifying, 5, 94, 176, 185, 305
igneous fusion, 291
imaginative, 284
imitative, 284
immersers: 253, 255, 258-259, 261, 266, 269, 282, 287, 307;
  creating, 284-286
immersing: 46, 72, 84, 88, 165, 199, 237, 225, 245, 258;
  techniques, 233, 287, 311
Immersing Experiences, 259, 262, 263
immersion, 194, 226, 243, 259, 266; personal, 226-220, 231
impulsive responders, 184, 187, 189
Indra's Net, 114
informing, 279, 296-297
infusion, 13-16, 36, 40-41, 47
insect repellents, 135-136
Institute for Earth Education (The), vi, 18, 27, 35, 44, 97, 99,
  101, 141, 145, 147, 157, 251, 261, 279, 281-282, 292, 296,
  310-313, 315
integrate, 13, 252, 307, 313
intellectual, 138, 152, 162, 175, 185, 258, 280, 284-285
interdependence, 10, 17, 98, 114, 125
interdisciplinary, 12, 15
internalizing, 18, 20, 132-137, 164, 249, 291, 296
Interpretive Loops, 263
interrelationships, 6, 103, 113-115, 119, 132, 214, 269, 271,
  276-278
intuitive, 95, 98, 258, 259

investigation: science, 7, 265; projects, 22, 66
Isle Royale, 241-242
issues, 12, 19-22, 47, 90, 157-158, 221, 252

j

j-curve, 108
journals: 202-206, 237-238, 241-242; inputs, 243-244
"Journal for Environmental Education," 14
Journey Home, 212
joy, 120-122, 125, 131, 138, 228-229, 236, 242, 288-289, 313

k

Keeper of the Magic, 299
Keepers of the Secrets, 299
Keller (Helen), 229
key concept statement, 211, 276; in magic, 200
KEYS (Earthkeepers), 256, 284
kinship, 120, 123-124, 131, 138

l

labeling (see naming), 169-175, 270-271, 306, 311
landmarks: names, 176-177; historical, 12
laws of energy, 105, 118-119, 123, 192
leadership, vii, 23, 31, 34, 67, 163, 281, 298-306
leaf, 106, 124, 140, 142, 211-212, 214, 274-275, 300-301
learning: adventure, 169, 197-204; mechanics, 61, 65-69, 72,
        80; model, 8, 15, 203, 279-280, 296; outcomes, 5, 7-8,
        15, 26-28, 41, 62; styles 26
learning programs: 5, 37-39; characteristics, 16-19
learning helper, 250-251
lecture method, 274
lessening impact (tasks), 147
life-changing, 18, 228, 307
lifestyles: changing, 3, 12, 17, 19-24, 31, 33, 97, 131, 159,
        162-164, 249; consumptive, 91, 101, 108, 128, 131, 152,
        156; examining, 6, 8, 33; gnome, xi; harmonious, 33,
        133, 144-149, 157, 162-163
Lincoln (Abraham), 182
litter (pick up), 155-16, 22
living in harmony, 145-146, 256, 264
living lightly, 14, 42, 197, 315

n

o

professionalism, 36-37
profile (living more lightly), 216-224
program building: 29, 45, 169, 250; steps, 253-306
programs: canned, 29-30; characteristics, 269-270; definition, 249-250; model earth education, 283
program specialists, 307
progress, 119, 151
Project Learning Tree, 34, 37-43
Project Wild, 34-35, 37-43, 207-209
Prop Manager, 299
props, 193, 196, 200, 203, 206, 212, 296, 302, 305, 311
pulling (vs. pushing), 67, 236
purpose, 253-254, 282
pyramid: earth education, 86-88, 100, 245; ecological, 107

q

questioning strategy, 181-192
questions: about activities, 293-294; and toilets, 110; twenty, 43, 169, 181-189, 192, 274, 300, 303; using, 187-188, 212, 304-305; "What Is It?", 175-179

r

rabbit, 106, 123
Rainbow Chips, 84-85
rangers, 200, 257
rationale, 253-254, 282
recreation, 5, 8, 24-28
recycling, 111, 147, 160, 219-220, 223
reflection, 216, 242-243, 264
reinforce: 139, 146, 169, 188, 205, 253, 266-267, 269, 272, 293-294, 299, 303-304, 306; techniques, 209-215
relate: 169, 253, 266-267, 274, 294, 306; techniques, 209-215
relating (Ways), 88, 238-245
relationship (sense of), 58, 63, 84, 94-96, 226, 231, 244, 262, 290, 310, 317
repetition, 205, 212
reproduction, 129, 145, 222, 224
resources: consumption, 12, 108; management, 91, 97, 125, 128; renewable, 217
reverence, 120, 124-127, 131, 138, 289
reward: 67, 69, 169, 188, 206, 208, 253, 266, 269, 294, 298, 305; techniques, 209-215
riddles, 200-201, 257, 286
roleplay, 206, 231, 234, 299
ruins, 260, 269

S

samurai, 203

Saving the Earth: A Citizens Guide to Environmental Action, 162

"S and S," 68, 261

schistosomiasis, 126

schools: 4, 9, 14, 23, 34, 159; curriculum, 15, 17-19, 39, 44, 235, 313; energy consumption, 22, 67-68; getting out of, 24, 26, 30, 42, 66, 228

Schweitzer (Albert), 124

science, x, 7, 17-18, 32, 37, 98, 100, 117, 148, 157, 276, 311

science education: 17-18, 85; investigation, 17, 265

scientific method, 17

secondary concepts, 6, 9-10, 47, 249, 252

secret ingredient, 72-82, 310

seeds, 106-108

Seeing Stick, 143-144

seeking patterns, 230

semantics, 13

sense of beauty: 232; in Tellurian gnomes, xi

sense of place, 232, 255, 265, 288

sense of relationship, 58, 63, 84, 94-96, 226, 231, 244, 262, 290, 310, 317

sense of time, 232

sense of wonder, 33, 228, 231, 255, 262, 265, 270, 283, 288, 297, 305

senses, 63-64, 72, 80, 83, 176, 201, 213, 228-234, 258-259, 264, 286

sensory awareness, 5, 63, 83, 95

Sessions (George), 99, 151

Seton (Ernest Thompson), 243

sense-watching, 58, 243

sewage treatment, 110

sharing: 203-209, 229, 244, 256, 263, 284; techniques, 206

sharing and doing, 100, 169, 204-209, 270, 306

Sharing Centres, 128

Sharing Circles, 206

sharpening senses, 228, 233, 264

Shepard (Paul), 150, 226

show and tell, 185, 204

Silent Spring, 123, 152

skills (environmental action), 158

Skinner (B.F.), 280

social ecology, 99, 150

social studies, 7, 17, 22-23, 25, 158

socialization experiences, 25-26

Socrates, 183

socratic method, 183

soil (see cycles), 70-71, 90-91, 105, 109, 112-113, 116, 121, 186-187

soil ceremony, 301, 304

t

u

V

W

Y

Z

...ATION . . . a new beginning